1,000,000 Books

are available to read at

Forgotten Books

www.ForgottenBooks.com

Read online
Download PDF
Purchase in print

ISBN 978-1-330-49340-3
PIBN 10069293

This book is a reproduction of an important historical work. Forgotten Books uses state-of-the-art technology to digitally reconstruct the work, preserving the original format whilst repairing imperfections present in the aged copy. In rare cases, an imperfection in the original, such as a blemish or missing page, may be replicated in our edition. We do, however, repair the vast majority of imperfections successfully; any imperfections that remain are intentionally left to preserve the state of such historical works.

Forgotten Books is a registered trademark of FB &c Ltd.
Copyright © 2018 FB &c Ltd.
FB &c Ltd, Dalton House, 60 Windsor Avenue, London, SW19 2RR.
Company number 08720141. Registered in England and Wales.

For support please visit www.forgottenbooks.com

1 MONTH OF FREE READING

at

www.ForgottenBooks.com

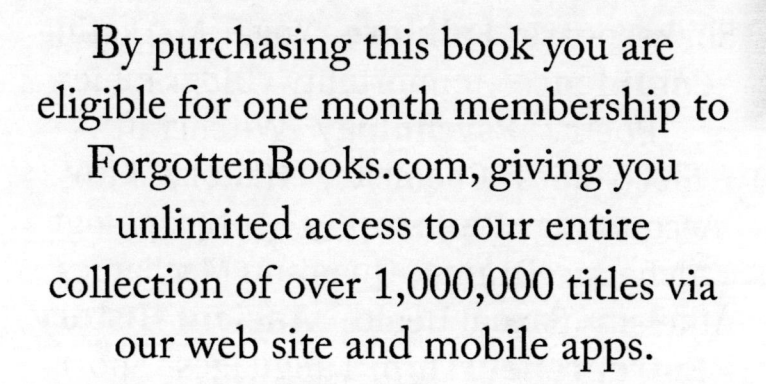

By purchasing this book you are eligible for one month membership to ForgottenBooks.com, giving you unlimited access to our entire collection of over 1,000,000 titles via our web site and mobile apps.

To claim your free month visit: www.forgottenbooks.com/free69293

* Offer is valid for 45 days from date of purchase. Terms and conditions apply.

English
Français
Deutsche
Italiano
Español
Português

www.forgottenbooks.com

Mythology Photography **Fiction**
Fishing Christianity **Art** Cooking
Essays Buddhism Freemasonry
Medicine **Biology** Music **Ancient Egypt** Evolution Carpentry Physics
Dance Geology **Mathematics** Fitness
Shakespeare **Folklore** Yoga Marketing
Confidence Immortality Biographies
Poetry **Psychology** Witchcraft
Electronics Chemistry History **Law**
Accounting **Philosophy** Anthropology
Alchemy Drama Quantum Mechanics
Atheism Sexual Health **Ancient History**
Entrepreneurship Languages Sport
Paleontology Needlework Islam
Metaphysics Investment Archaeology
Parenting Statistics Criminology
Motivational

THE
SCENERY OF SCOTLAND

VIEWED IN CONNEXION WITH ITS PHYSICAL GEOLOGY.

BY

ARCHIBALD GEIKIE, F.R.S.

FELLOW OF THE ROYAL SOCIETY OF EDINBURGH; OF THE GEOLOGICAL SOCIETY OF LONDON; AND OF THE GEOLOGICAL SURVEY OF GREAT BRITAIN.

WITH A GEOLOGICAL MAP BY SIR RODERICK I. MURCHISON, K.C.B. F.R.S. AND ARCHIBALD GEIKIE, F.R.S. EDIN AND ILLUSTRATIONS.

London and Cambridge:
MACMILLAN AND CO.
1865.

LONDON:
R. CLAY, SON, AND TAYLOR, PRINTERS,
BREAD STREET HILL.

TO

SIR RODERICK I. MURCHISON,

K.C.B. F.R.S.

Director-General of the Geological Survey of Great Britain and Ireland.

MY DEAR SIR RODERICK,

To no one can a new volume on Scottish Geology be more fitly inscribed than to you, to whom Scottish Geology owes so much; nor perhaps can such an inscription come with greater appropriateness than from one who has worked with you, hammer in hand, over many a league of Highland ground, and who is further bound to you by many acts of courtesy and kindness. You are of course in no way responsible for the opinions expressed in the following chapters. That they may sometimes be at variance with your well-known views of the same subjects would only serve, if need were, as an additional motive for offering to you this expression of respect and esteem.

Believe me to remain,

Yours very truly,

ARCHIBALD GEIKIE.

GEOLOGICAL SURVEY, AYR.
    2d *June*, 1865.

PREFACE.

To trace back, if that might be, the origin of the present surface of the country, and by working out the structure of the rocks, to contrast the aspect of the land to-day with its condition in former geological periods, has been to the author of these pages the delightful occupation of years. The writing of this volume has thus gone on side by side with daily labour in the field, amid all the changes of scene and surroundings that fall to the lot of a member of the Geological Survey. Some portions have already appeared in the *North British Review* and elsewhere; but these were written with the view of ultimate reunion and republication in some such form as that in which they now appear.

The principles which have guided me in the following investigation are far from new: they were laid down long ago by Hutton and Playfair, and they have recently received fresh illustration from the pen of Professor Ramsay. I can claim nothing more than to have tried in some detail to develop these principles in an inquiry into the origin of the existing scenery of Scotland. The views to which I have been led, however, run directly counter to what are still the prevailing impressions on this subject, and I am therefore prepared to find them disputed, or perhaps thrown aside as mere dreaming. That in searching for a pathway through a field of scientific research wherein the travellers have as yet been few, one can hardly fail to go here and there astray, that he must needs miss much by the way, and that a few steps to either side would sometimes have brought him out of the cloud of doubt and uncertainty through which at the time no outlet could be seen, will by none be more frankly admitted than by the traveller himself. Yet in spite of these mishaps, he may believe that on the whole his journey has been a progress in the right direction, and that even his errors may not be without their use in pointing out the right track to other explorers. Such at least is the hope in which I lay these chapters before my brethren of the hammer.

No one may venture to write on the subjects discussed in this volume without first studying what has been written upon them by Professor Ramsay. To his papers, but still more to many long discussions with him by the fireside and among the hills, and to the stimulus from contact with a mind so original in conception, and so marked for its faculty of grasping at once the details and the broad bearings of a question, it is a pleasure to express my obligations.

While the following chapters were passing through the press, a very ingenious and original work by Mr. J. F. Campbell has appeared, with the title of *Frost and Fire.* He refers the glaciation of Scotland and other countries not to land-ice, as is held by Agassiz, Ramsay, Lyell, Chambers, and Jamieson, but mainly to icebergs and icefloes drifting from the north in ocean-currents. While he is thus opposed to my views, I am glad of this early opportunity of acknowledging the markedly courteous terms of his allusions to myself. There is much that is suggestive in his volumes; he has worked out the theory and proofs of berg-action in a way in which it has never been done before. He has not, however, refuted the proofs of the former wide extension of land-ice in this country. These still remain, it seems to me, as clear and decisive as ever. As this is no place for discussion, I will only add that bergs and icefloes, instead of scooping out valleys, would rather, I think, when their long-continued action is looked at broadly, tend to plane down the inequalities of the sea-bed. In reading *Frost and Fire* (as yet too hurriedly) it has occurred to me that what we call a plain of marine denudation may perhaps not always be the result of wave-action, but may have been more or less ground smooth by the yearly grating of those ice-rafts whose movements Mr. Campbell has so graphically described.

The accompanying Geological Map was first

published in 1861 with a pamphlet of Explanatory Notes. Sir Roderick Murchison and the Messrs. Johnston having readily given their consent to its being used in the present form, I have revised it with some care, so as to bring it up to the state of the science at the close of last year. The plates are mere pen and ink outline sketches which I have copied from my note-books upon transfer-paper. Owing, however, to my inexperience in lithography, they have sadly failed in the printing, and are very far from what I had hoped they would be. The woodcuts were drawn by myself on the blocks, and are likewise only copies of the rough sketches made on the spot.

CONTENTS.

CHAPTER I.
POPULAR IDEAS REGARDING THE ORIGIN OF MOUNTAINS AND VALLEYS 1

CHAPTER II.
POWER OF RAIN, RIVERS, SPRINGS, AND FROST UPON THE SURFACE OF THE EARTH. 14

CHAPTER III.
ACTION OF THE SEA AND OF WIND 39

CHAPTER IV.
ACTION OF GLACIERS AND ICEBERGS 77

I. THE SCENERY OF THE HIGHLANDS.

CHAPTER V.
GEOLOGICAL STRUCTURE OF THE HIGHLANDS—CONNEXION OF GEOLOGICAL STRUCTURE AND SCENERY—ANCIENT TABLE-LAND OF THE HIGHLANDS 91

CHAPTER VI.
ORIGIN OF THE HIGHLAND VALLEYS — VALLEYS CROSSING WATERSHEDS; PASSES OR BEALLOCHS—SEA-LOCHS—VALLEY-SYSTEMS OF THE HIGHLANDS 114

CHAPTER VII.

INFLUENCE OF ANCIENT GLACIERS AND ICEBERGS ON HIGHLAND SCENERY—TRACK OF THE FIRST GREAT ICE-SHEET OF THE GLACIAL PERIOD—POSSIBLE THICKNESS OF THE ICE—ORIGIN OF HIGHLAND LAKES—DÉBRIS MADE BY THE ICE-SHEET—HIGHER ELEVATION AND SUBSEQUENT SUBMERGENCE OF THE HIGHLANDS—SECOND GLACIER-PERIOD IN THE HIGHLANDS 149

CHAPTER VIII.

CAUSE OF THE LOCAL VARIETIES OF HIGHLAND SCENERY . . 207

II. THE SCENERY OF THE SOUTHERN UPLANDS.

CHAPTER IX.

THE SOUTHERN UPLANDS—GEOLOGY OF THE DISTRICT—CHARACTER OF THE SCENERY—SURFACE OF THESE UPLANDS AN OLD SEA-BOTTOM—VALLEYS—INFLUENCE OF GLACIERS AND ICEBERGS 231

III. THE SCENERY OF THE MIDLAND VALLEY.

CHAPTER X.

THE MIDLAND VALLEY—GEOLOGICAL STRUCTURE—DENUDATION 266

CHAPTER XI.

CAUSE OF LOCAL VARIETIES OF LOWLAND SCENERY—TRACES OF GLACIERS AND ICEBERGS 292

CHAPTER XII.

CHANGES IN THE SCENERY OF SCOTLAND SINCE THE ADVENT OF MAN—RAISED BEACHES—DISAPPEARANCE OF THE ANCIENT FORESTS; GROWTH OF PEAT-MOSSES—INFLUENCE OF MAN UPON THE SCENERY OF THE COUNTRY 319

CHAPTER XIII.

RECAPITULATION AND CONCLUSION. 340

DIRECTIONS FOR BINDER.

Coloured Map to face Title Page.

PLATE I.	. . . . to face	154
,, II.	. . . .	210
,, III.	. . . .	218
,, IV.	. . . .	224
,, V.	. . . .	262
,, VI	. . . .	296

To roam at large among unpeopled glens
And mountainous retirements, only trod
By devious footsteps; regions consecrate
To oldest time! and while the mists
Flying, and rainy vapours, call out shapes
And phantoms from the crags and solid earth;
. . . . . . and while the streams
Descending from the region of the clouds,
And starting from the hollows of the earth
More multitudinous every moment, rend
Their way before them—what a joy to roam
An equal among mightiest energies!

WORDSWORTH, *Excursion*, b. iv.

SCENERY OF SCOTLAND.

CHAPTER I.

POPULAR IDEAS REGARDING THE ORIGIN OF MOUNTAINS AND VALLEYS.

AMONG the many forms of scenery which vary the surface of the earth, none fixes itself more firmly on the memory and the imagination than that of the hills and mountains. The aspect of the lower grounds is liable to constant change. We see the waves wearing down the shore, and the rivers ever and anon devastating the meadows. Man himself does much to transform the landscape: he ploughs up peat-mosses, cuts down forests, plants new woodlands along the hill-sides, covers the valleys with cornfields and orchards, graves the country with lines of roadway, and builds his cottages, villages, seaports, and towns. But high above the din and stir of these changes, the great mountains rise before us with still the same forms of peak and crag that were familiar to our forefathers long centuries ago. Amid the fleeting outlines of the lowlands these alone seem to defy the hand of time. And hence "the everlasting hills" and "ancient mountains" have ever been favourite emblems of permanence and grandeur.

Not that signs of revolution failed to be perceived in the early ages among the crags and valleys of the hills. There is an air of ruin and waste about these high grounds which no eye can miss. But the cleaving of their ravines, the scarping of their precipices, the opening of their valleys, and the strewing of their slopes with piles of loose rock, were looked upon vaguely as the complex result of the first grand upheaval of the mountains out of chaos.[1]

---

[1] This popular belief, which even yet, in spite of modern science, has not become extinct in this country, is well expressed by Milton:

> "When God said,
> 'Be gathered now, ye waters, under Heaven
> Into one place, and let dry land appear;'
> Immediately the mountains huge appear
> Emergent, and their broad bare backs upheave
> Into the clouds! their tops ascend the sky:
> So high as heav'd the tumid hills, so low
> Down sunk a hollow bottom, broad and deep,
> Capacious bed of waters."—*Paradise Lost*, Book VII.

Sometimes, however, the traces of destruction were too marked to be assigned even by the popular mind to the first creation of all things. The hills looked as though, long after their birth, they had been rent in twain; the cliffs were shattered and broken as if they too had suffered from a like catastrophe; the huge fragments of crag and mountain scattered over the declivities, or lying thickly in the valleys below, seemed all to tell of some conflict, later, indeed, than the making of the world, yet lost in an antiquity far beyond the records of man. To the influence of scenery of this kind on the mind of a people, at once observant and imaginative, such legends as that of the Titans should in all likelihood be ascribed. It would be interesting to trace back these legends to their cradle, and to mark how much they owe to the character of the scenery amongst which they took their rise. Perhaps it would be found that the rugged outlines of the Bœotian hills had no small share in the framing of Hesiod's graphic story of that primeval warfare wherein the combatants fought with huge rocks, which, darkening the air as they flew, at last buried the discomfited Titans deep beneath the surface of the land. Nor would it be difficult to trace a close connexion between the present scenery of our own country and some of the time-honoured traditionary stories of giants and hero kings, warlocks and witches, or between the doings of the Scandinavian Hrimthursar, or Frost-Giants, and the more characteristic features of the landscapes and climate of the north.

But apart from the region of fable and romance, it is impossible to wander among the glens and solitudes of a wild mountainous tract without feeling a certain vague awe, not merely on account of the magnitude or loneliness of the surrounding scenery, but from the mystery that seems to hang over its origin. The gentle undulations of a lowland landscape may never start in the mind a passing thought; but we are arrested at once by the stern broken features of the hills, and cannot help asking ourselves how they came into being.

To such a question the natural answer is sought in vast primeval convulsions, that suddenly tossed up the mountains, rent open ravines and glens for the rivers, and unfolded wide valleys to receive and remove the drainage of the country. There is an air at once of simplicity and of grandeur about this explanation which has made it a favourite and popular one. It deals with that dreamland of conjecture and speculation lying far beyond the pathways of science, where one has no need of facts for either the foundation or superstructure of his theory. It thus requires no scientific knowledge or training; it can be appreciated by all, and may be applied to the history of a mountain-chain by one to whom the very name of geology is unknown. No man, indeed, who is aware of what has been ascertained of the history of the earth can hold this popular notion to the full. For it is now well known that the mountains are of many different ages, that some of them were rent and worn down before others had begun to be, and that the rocks which form the present surface of the earth are the result of many successive ages of geological change. But even where this is familiar knowledge, the vaguest and most indefinite ideas prevail regarding the origin of the scenery which now surrounds us. It seems, indeed, not a little curious, that notwithstanding the amount of geological knowledge diffused through this country regarding the origin of the various systems and formations which lie beneath the surface, so much ignorance and uncertainty should exist with respect to the origin of the surface itself. We know much regarding the history of the Silurian grits and limestones, of the Old Red sandstones and conglomerates, of the Carboniferous shales and coals, but when we are asked to tell how these rocks came to be moulded into the outlines which they wear at the surface, and thus how our present scenery has arisen, we cast about for an answer, and learn that it is not easy to find. And the more the question is pondered over the less obvious is the answer seen to be.

The following chapters are an attempt to find a

solution of the problem, in so far as it relates to the history of the scenery of Scotland. That this special application may be made the more intelligible to non-geological readers, it may be well to begin by stating as briefly as possible the terms of the problem to be solved, and, after pointing out the inadequacy of the common explanation, to direct the notice of the reader to the nature and results of those processes whereby the present surface of the globe is slowly but ceaselessly modified. Having thus considered how scenery is changing at the present day, we shall be the better able to understand how changes of a like kind have been brought about in the past, and how their accumulated effects may have resulted in that varied contour which now diversifies the country.

To whatever source the origin of the existing configuration of the surface is to be traced, the result of the whole has been a system of the most nicely-adjusted symmetry. Hill and mountain, valley and glen seem each as if made to fit in closely with its neighbours. One main office of land is to serve as a channel on which a part of the water which rises from the sea into the clouds may find its way once more to the deep. The manner in which this task is accomplished, familiar and even common-place though it may be, can hardly be thoughtfully contemplated without wonder. From the high grounds the gathered rains and springs descend in hundreds and thousands of watercourses, which, from the tiniest runnel up to the ample river, are all arranged in the closest harmony with each other, and with the whole orderly system of which they form a part. This well-balanced symmetry cannot but have resulted from some general cause, acting uniformly throughout the land which it fashioned for its own ends.

The form of the valleys may vary indefinitely without disturbing the general symmetry. They may be wide, open, smooth, with gently shelving sides, or they may be only narrow gorges in which the rivers toil between naked walls of rock. Merely from the map one cannot tell where the valleys are wide and where narrow. The most precipitous ravines fall easily into the general plan, and lie in the paths of the streams just as naturally and unconstrainedly as do the widest straths. Nor should we fail to remark that the ground-plan of the valleys, as defined on a map by the courses of the streams, is marked everywhere by the same great features, whether the region be one of mountains or low plains. If the map has no hill-shading it may be impossible to tell which are the higher mountain ranges, and which the lower groups of hills and plains, so uniform is the disposition of the valleys and water-courses apart from relative height.

We have to inquire by what probable means this

harmonious grouping of hill, and valley, and watercourse, was brought about.

Two explanations have been given. One calls in the aid of old terrestrial convulsions, and looks on the valleys and ravines as due to fractures and subsidences or upheavals of the earth's crust; the other, while far from ignoring the influence of subterranean movements, holds that these have not been the chief cause in determining the present form of the ground, but that the valleys are due in the main to *denudation*, or the erosive action of rains, streams, ice, and other subaerial forces. Of these two opinions the latter alone appears to me tenable with regard to the hills and valleys of Scotland, and as the following chapters are founded upon it, there may be some advantage in stating here in a few words the grounds for the belief.

It should be borne in mind that in dealing with the past history of our planet the element of *time* ought never to be lost sight of. To turn away from the known and visible causes of change, and to imagine the former presence of others far more stupendous, because the results achieved seem otherwise inexplicable, is to substitute mere speculation for inductive reasoning. In all such attempts we make the fatal error of forgetting that in the geological history of our globe *time is power*. It may not be easy to get thoroughly rid of the natural tendency to associate great effects with great causes, but if one can free himself from this prejudice he will probably be led, in the spirit of Hutton, to perceive that there appears to be no great change in the configuration of the land which may not have been brought about by the working of those slow every-day processes which are in progress now.

There can be no dispute regarding the abundance of the upheavals, subsidences, and dislocations which the crust of the earth has undergone. But that our valleys and ravines are not mere cracks, would seem to be put beyond dispute by the fact that for one valley which happens to run along the line of a dislocation, there are, I dare say, fifty or a hundred which do not.[1] Moreover, it can be shown that out of every valley and glen a great mass of solid rock has been carried bodily away, and that even the highest mountain-tops have suffered a similar loss. If we could restore the missing material we should in truth be able to fill up the glens and valleys again, so that the mountainous parts of the country would thus be turned into rolling table-lands.

But perhaps the most evident argument against the doctrine of fracture and convulsion, and in favour of the Huttonian theory of erosion, is to be found in the

---

[1] There is no point which the detailed investigations of the Geological Survey have made clearer than this.

very grouping of the valleys themselves. It appears to me hard to see how a thoughtful survey of the configuration of a land-surface can lead to any other conclusion than that "the mountains have been formed by the hollowing out of the valleys, and the valleys have been hollowed out by the attrition of hard materials coming from the mountains."[1] The independence of each hydrographical basin, the nice adjustment of all its parts, the union of the different basins in one great system, and the constant reference of the whole to the grand end of conveying the drainage from mountain-top to sea-shore, point not to the random outbreaks of underground violence, but to some general orderly agency, which has shown a wonderful harmony and consistency in every part of its working. What power save that of running-water and of the other forms of atmospheric waste, could have achieved these results? Or where else may we look for a force that could so accurately carve out the valleys for the requirements of the streams, unless in the action of the rains and streams themselves? "If, indeed," says Playfair, "a river consisted of a single stream, without branches, running in a straight valley, it might be supposed that some great concussion or some powerful torrent had opened at once the channel by which its waters are conducted to the ocean; but when the usual form of a river is considered,

[1] Hutton's Theory of the Earth, vol. ii. p. 401.

the trunk divided into many branches, which rise at a great distance from one another, and these again subdivided into an infinity of smaller ramifications, it becomes strongly impressed upon the mind, that all these channels have been cut by the waters themselves, that they have been slowly dug out by the washing and erosion of the land; and that it is by the repeated touches of the same instrument that this curious assemblage of lines has been engraved so deeply on the surface of the globe."[1]

Did the reader ever stand on a flat shore and watch how the water which had soaked into the sand just below the upper limit of the tide trickles down the seaward slope towards the pools and shallows on the lower part of the beach? He could hardly find a better illustration of the drainage of a country. The water that oozes out from below high-tide mark begins by degrees to gather into tiny runnels; these gain size and speed as they descend, often flowing into each other, and thus with their united torrent cutting narrow and sometimes tortuous channels for themselves out of the sand. If the locality be a favourable one, these miniature rivers

---

[1] Illustrations of the Huttonian Theory, § 99. If the reader has not already read this classic, let me strongly urge him to do so. He will there see how far Hutton and Playfair were ahead of many modern geologists in the breadth of view they took of physical geology. And he will find their opinions set forth with a terseness, clearness, and elegance, such as may be found in no other English treatise on physical geology.

may be seen undermining their banks, and sweeping the *débris* away to sea. Thus the sand which wore, perhaps, a perfectly smooth surface when the tide left it a few hours before, is now channelled and worn into diminutive valleys, gorges, and ravines, with narrow ridges and broader *plateaux* between them. It might then be taken as a kind of relief model of the drainage of one side of a country. As the process of erosion goes on, the likeness of the beach to a series of river-systems grows every minute more marked. But at last the turned tide comes back and levels the whole, thus illustrating what geologists call "a plain of marine denudation." Yet again this levelled surface, when the tide retires, is once more exposed, the same system of water-carving goes on as before, and a new system of valleys, ravines, water-courses, ridges, and table-lands makes its appearance.[1]

Now it is, I believe, in this kind of way that a great river-system is excavated. The process is then of course an infinitely longer one, calling in, as we shall see, the agency of rain, springs, streams, and ice, and making these all work together for the accomplishment of the general end. But in either case the ultimate result is achieved by denudation.

---

[1] Since writing the above, I find that a similar illustration has occurred to my colleague, Mr. Jukes. "Student's Manual of Geology," 2d edit. p. 105.

Water seeking its way seaward cuts a network of paths for itself; an hour or two is enough to channel the sandy beach, millions of years may be needed to cut down a mass of high ground into mountain and glen; but in the long lapse of geological time the one result is doubtless as sure as the other.

The conclusion, therefore, to which an attentive examination of the present surface of the country points is, that although the rocks have unquestionably suffered much from subterranean commotions, it is not to that cause that their present external forms are chiefly to be traced; that the mountains exist, not because they have been upheaved as such above the valleys, but because their flanks having been deeply cut away they have been left standing out in relief; and that the valleys are there, not by virtue of old rents and subsidences, but because moving water, with its help-mates frost and ice, has carved them out of the solid rock. These principles will be further developed in succeeding chapters, when they are applied to the elucidation of the history of Scottish scenery. In the meantime let us proceed to mark how the various denuding forces, which are held to have done so much in the past, are now carrying on their work around us.

CHAPTER II.

POWER OF RAIN, RIVERS, SPRINGS, AND FROST, UPON THE SURFACE OF THE EARTH.

SOME hesitation is not unnaturally felt in admitting that such marvellous results as the scooping out of our valleys and ravines can have been brought about by any mere surface-action. It may be well, therefore, to look for a little at the nature of the different processes which are now modifying the surface, and to which the origin of the existing scenery is sought to be referred.

In such an inquiry we are soon struck by the important part which water plays in these changes. There is a ceaseless passage of water from land to sea, and from sea back again to land.[1] In the latter half of the cycle clouds and winds are the vehicles that carry the moisture, but in the former half,

---

[1] There is more scientific truth than most readers imagine in Homer's reference of all waters to the parentage of the great Ocean:

Βαθυρρείταο μέγα σθένος Ὠκεανοῖο,
Ἐξ οὗπερ πάντες ποταμοὶ καὶ πᾶσα θάλασσα
Καὶ πᾶσαι κρῆναι, καὶ φρείατα μακρὰ νάουσιν.—*Il.* XXI. 195.

where the water finds its way from the land to the sea, it is conducted in innumerable channels worn into the surface of the earth. When it falls to the ground it is nearly pure, but when after its journey from mountain to shore, it returns to the sea, it is charged with mineral admixture, partly in chemical solution and partly in mechanical suspension. As these ingredients are of course derived from the rocks and soils over which the gathered moisture has flowed, it is plain that a constant waste is thus carried on, and that the spoils of the land are borne ceaselessly downwards to the ocean. Such is one of the first features to arrest the attention; and even before entering into the details of this process of erosion, the inquirer, on reflection, will be led to perceive that since the waste is universal it must have worked vast changes on the surface of the globe, and that, indeed, if allowed time enough, there may be no limit to its operations until the land is everywhere wholly worne away.

Rain.—The rain which falls on the land partly runs off again at once in brooks and rivers, partly sinks beneath the surface and reappears in springs. Each part has its own share in the general erosion of the rocks. It has been customary, in enumerating the various geological agents, either to pass rain over, or to assign to it a very subordinate place. But it will perhaps be found in the end that no other power has done more to modify the surface of the land. From the very tardiness and apparent insignificance of its results, geologists have been apt to overlook its action. It cannot be doubted, however, that even before the rain collects into brooks it effects some change, especially when flowing over soil or decomposing rock. This is often well shown after a heavy shower has fallen upon a piece of newly-ploughed sloping land. The finer earth or mud is then found collected in pools between the furrows at the bottom of the field, and, as the result of ages of such washings, the thickest and richest loam is spread along the foot of the slope. Were it not that the soil on the upper part of the field is always being renewed by the decomposition of the stones and the growth of vegetation, the surface there would be in the end swept bare of soil altogether. The rain which washes down the finer particles to the bottom of the declivity likewise carries a portion off to the nearest brook, and thus prevents the indefinite and unchecked accumulation of loam on the lower grounds. In the south-east of England, where, as the land did not sink beneath the sea during the glacial period, it has been exposed for a very long period to the uninterrupted action of the atmosphere, the results of this gentle transporting power of rain are visible in thick accumulations of *brick-earth*, a fine loam or earth, lying on the valleys and lower grounds, and derived from the adjoining slopes.[1] In Scotland it is only the post-glacial *rain-wash* that is now to be seen. But this is often several feet thick, and spreads out in patches over the older formations. The effects of rain in washing off the lighter soil from a hill-side may often be well seen among the uplands of the southern counties, where the original covering of boulder-clay has been by degrees carried away from the smooth summits and higher parts of the slopes, but the striated stones of the deposit, being heavier and not so easily removable, still often lie pretty thickly among the turf up even to the hill-tops.

Such facts show that, apart from the action of streams, rain exerts an appreciable power of abrasion on the surface of the land. Its influence is gentle and slow, and on that account apt to be overlooked. But with time enough given for its work, we can hardly place a limit to its results. For rain, like the other agencies of waste, does not act alone. It is in close league with chemical decomposition, and the destructive power of frosts, springs, brooks, and rivers.

[1] The proofs of the vast influence of rain in modifying the surface of the south-east of England, have been carefully studied by my colleagues of the Geological Survey, Mr. Foster and Mr. Topley, who have been led to the conclusion that the well-known denudation of the Weald has been effected by sub-aërial action—a deduction in which they agree with Professor Ramsay ("The Physical Geography and Geology of the British Isles," Lect. III.) They are at present engaged in Preparing a paper on this subject for the Geological Society.

Whether it flows over solid rock or loose earth, it is constantly washing off the lighter parts, leaving the rest to be broken up by exposure to the air, and eventually swept away in turn. Thus every year must imperceptibly lower the general level of the land. In dealing with the vast cycles of geological time, we are likely rather to take too low than too high an estimate of the potency of this quiet agency in modelling the present contour of the country.

Brooks and Rivers.—When rain falls it quickly flows into the minor hollows of the surface, gathering there into little runnels that unite into brooks, thence into larger streams,[1] and finally into rivers. An examination of these watercourses will show the evidence of the assertion of Hutton, that they are in general the work of the water which flows in them. Beginning at the hill-tops we first meet with the spring or "well-eye," from which the river takes its rise. A patch of bright green mottling the brown heathy slope shows where the water comes to the surface, a treacherous covering of verdure often concealing a deep pool beneath. From this source the rivulet trickles along the grass and heath, which it soon cuts through, reaching the black, peaty layer

---

[1] Such streams, intermediate in size between brooks and rivers, are known in Scotland as "waters;" for example, the Water of Girvan, the Water of Cree, Water of Leith, Fassney Water, Dye Water, &c. They usually flow into rivers, of which they are then the chief feeders, but some enter the sea.

below, and running in it for a short way as in a gutter. Excavating its channel in the peat, it comes down to the soil, often a stony earth bleached white by the peat. Deepening and widening the channel as it gathers force with the increasing slope, the water digs into the coating of drift or loose decomposed rock that covers the hill-side. In favourable localities a narrow precipitous gully, twenty or thirty feet deep, may thus be scooped out in the course of a few years. Its rapid growth may often be impressively seen at its upper end. There the stream, flowing in its narrow gutter through the peat, suddenly descends into the gully in a waterfall that tumbles over a vertical face of the superficial detritus. The dash of the spray and the percolation of water are ever loosening large masses both from the sides of the gully and from behind the waterfall. Blocks of peat several yards in circumference still attached to the underlying soil lie in tumbled ruin at the bottom of the ravine, where they are eventually broken up and washed away down to the valleys. The waterfall is thus creeping up the hill, and the gully is becoming longer as well as wider and deeper. Such a scene as this may often be observed in sheep-drains incautiously made on steep slopes. The drains, originally cut, perhaps, merely in the peat, have become the channels of torrents during a rainy season. They have thus been torn up, and turned into long yawning chasms, which every winter dig deeper into the side of the hill. The higher slopes of the hills and mountains are in this way seamed with narrow gullies, cut either by rains or by the outflow of springs, sometimes in a covering of superficial *débris*, sometimes in the rock itself, and the detritus so excavated is spread out in a fan-shape at the edge of the valley below.

> "Thus oft both slope and hill are torn,
> Where wintry torrents down have borne,
> And heaped upon the cumbered land
> Its wreck of gravel, rocks, and sand."

Following one of these streamlets to the foot of the hill, let us trace its course as it enters a valley where, joined by tiny tributaries from either side, it slowly gains in size. For a time its path winds from side to side in endless curves, which sometimes nearly meet and enclose little flat islets of meadow. Listlessly and smoothly it glides along, until after a mile or two its channel becomes deeper and more rocky. The streamlet chafing within its narrowed bounds enters a thick copsewood, in which we hear it murmuring. Then comes the roar of a waterfall, beyond which there is silence. Breaking through the thick wood some way below the waterfall, we find ourselves on the edge of a deep ravine, and yonder at the bottom is our old friend the streamlet once more as placid as among the reeds and meadows of the upper valley. The long pendant willow-branches that just kiss the surface of the water are scarcely shaken, while the polished pebbles that strew the channel lie unmoved by the rippling current that glides over them. How, we involuntarily ask, came this deep ravine into existence?[1] As we looked down the valley from above, and saw the green wood along its bottom, we never imagined that it concealed so deep a gash as this. Our eye scans the precipitous walls of the dell, with their rocks cleft through to a depth of perchance fifty feet. It requires no great scrutiny to perceive that the beds of stone on the one side form the prolongations of those on the other, and that consequently there must have been a time, ere yet the ravine existed, when these beds stretched along unbroken. Satisfied with this conclusion, our first impulse may be to bethink us of some primeval earthquake, when the solid land was broken up by great rents and yawning chasms. Into one of these clefts we might suppose the little streamlet to have eventually found its way, moistening the bare, raw, barren rocks, until at last their surface put on a livery of moss, or lichen, or liverwort, and the birch, the alder, and the willow found a nestling-place in their crevices.

[1] The following illustration of river-action is partly taken, with some verbal changes and additions, from my little volume, "The Story of a Boulder," p. 162.

Such a view of the origin of the wooded dell is certainly that which instinctively suggests itself. We know that earthquakes break open the ground into hideous chasms: we do not in the brief experience of a single life-time see such chasms formed by any other agency; consequently, it appears a legitimate conclusion to infer that such a ravine as that now described can only have been produced by a subterranean convulsion. Let us descend into the dell, and see whether there remains any evidence of such a dislocation.

Standing among the stones in mid-channel, we soon find that the floor on which the stream is flowing consists of sandstone and shale, and that the walls of the little gorge are formed of similar materials. Fixing on one of the sandstone beds which has been hollowed into pools and channels by the running water, we trace it across to the left-hand side of the ravine, and away up into the precipitous cliff, till it is lost amid the ferns and brushwood. There can be no doubt, therefore, that the ledge on which we were but now standing is a continuous portion of the rocks that form the left side of the ravine. Returning again to the centre of the stream, we proceed to trace out the course of the other end of the same sandstone bed, and find that it, too, strikes across to the rocks on the right-hand side without a break or fissure, and passes up into the cliff of which it forms a part. Clearly, then, this bed runs in an unbroken, unfissured line, from the one side to the other, and the rocks of either cliff form one continuous series. But lest by any chance there may be some part of the section concealed at this particular point of the dell, we pick

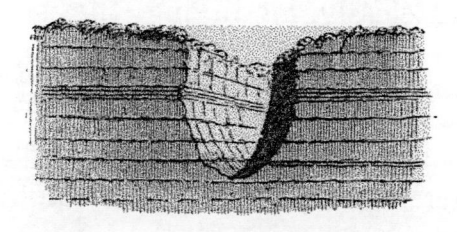

FIG. 1.—SECTION OF A RAVINE EXCAVATED BY A RIVER.

our way along the bottom for some two or three hundred yards higher up, and mark everywhere the same continuity of the strata from side to side, without a trace of any crack or fault. It is plain, therefore, that the ravine does not owe its existence to a subterranean dislocation, for in that case the rocks would show the break. What then? Whither shall we turn, it may be asked, to find another agency equally grand and powerful in its working and mighty in its results?

It is this habit of dwelling not on the nature of the process, but on the magnitude of the results of terrestrial change, which leads most men to view these river gorges as the work of some single, potent, and transient force. The completed change is before us in all its details, and the mind naturally associates this unity of effect with one powerful cause. It is only after reflection and study that one is led to perceive how inadequate such an interpretation may be, since it takes no heed of the element of time and the slow working of the many little agencies that are ceaselessly modifying the surface of the globe.

Returning again to the streamlet, let us continue the ascent of the channel, and by marking the changes in the character and features of the stone that forms the cliff on either hand, we may perhaps learn something as to the origin of the ravine. We come to a bare face of rock, where, as brushwood and herbage find but a scanty footing, the strata can be attentively studied. The cliff shows a number of beds of pale grey sandstone, alternating with courses of a dark, crumbling shale. The sandstones being harder and firmer in texture stand out in prominent relief, while the shales between have mouldered away, covering the bottom of the bank with loose shivers. We can mark, too, that as this decay goes on the harder beds continually lose their support, cracking across, chiefly along the lines of joint, and rolling down in huge angular blocks into the stream. Every year must needs add to this waste, and thus slowly widen the dell. For the broken fragments do not form in un--disturbed heaps over the solid rock below so as to protect it from the weather; they are evidently worn away by the stream, and their detritus is carried down the ravine onwards to the sea. More special reference will be made a few pages further on to the causes of this slow rotting-away of solid rocks: it is at present enough to note that it is always going on. The rubbish thus produced, whether mere sand or large blocks of stone, is borne away by the running water, so that the rocks of the ravine, instead of getting at last protection under a thick cover of their own ruins, are ever being exposed anew to the wear and tear of the elements. Nay, their very *débris* is used as a powerful instrument in further increasing their erosion. For every fragment of stone which is driven onward by the current acts as a file on the rocks along the sides or bottom of the channel. It gets worn down itself, but at the same time it helps to grind away the solid rock over which it is rolled. The process is indeed an infinitely slow one. During a short visit we of course cannot see any change actually accomplished, nor even if we were to return after the lapse of a generation might we be able to detect any appreciable difference. But each successive stage in the progress of the waste is illustrated before us, and the evidence is not less convincing than if we could follow the history of each block of stone from the time when loosened by springs or frosts, it fell from the cliff into the stream, down to the time when, after a long rubbing and grinding on the rocks of the watercourse, it is at last reduced to mere sand. Nevertheless, it may still be objected that this explanation merely accounts for the widening of a channel already formed, and affords no answer to the question how the ravine was formed at first.

We continue the ascent for a short way further. A scrambling walk through briars and hazel-bushes, sometimes on rocky ledges, high among the cliffs, sometimes among prostrate blocks that dam up the stream, brings us at last full in front of the waterfall whose sound we had heard in coming down the valley —a sparkling sheet of water that dashes over a precipitous face of rock some twenty feet high. The appearances observable here deserve a careful attention. Our eyes have not been long employed noting the more picturesque features of the scene ere they discover that the dark brown band of rock forming the summit of the ledge over which the water tumbles is continuous all round the sides of the dell. There is consequently no break or dislocation here. Approaching the cascade, we mark the rock behind it so hollowed out that its upper bars project several feet beyond the under ones. In this way the body of water is shot clear over the top of the cliff, without touching rock till it comes splashing down among the blocks in the channel. The surface of the rock in this recess is never dry; the spray of the fall constantly striking on it keeps it always dank and dripping. And thus this part of the sandstone tends to crumble away even more speedily than the ledge over which the water falls. Much depends on the texture and structure of the rock; sometimes the upper and sometimes the under part of the cliff will decay more rapidly, and thus the form of the waterfall must be ever changing. We can see, for instance, how the waste of the recess may go on until the upper projecting part of the cliff loses its support, and sinks with a crash into the rocky pool below, thus, perhaps, transforming the single cascade into a series of little falls, which tumble over the new surfaces of rock now exposed to the wear and tear of the stream. If, as the waste goes on, it should chance that at last a stratum of soft material, such as a bed of shale, came to form the lower part of the cliff, the rate of destruction would be greatly increased. If, on the other hand, a hard sandstone or limestone should occupy the same position, the rate would be in like manner diminished. But whether more slowly or more speedily, the waterfall is constantly cutting its way backward. Ages ago it stood many yards further down the ravine than it does now, the intervening space having been hollowed out by the recession of the fall. And ages hence it will be found much higher up. Thus the cascade recedes, and the ravine is dug deep out of the solid rock. And to this tardy but certain process is the origin of the long, narrow river-gorge to be ascribed.

We may learn yet a farther lesson by examining the channel of the stream above the fall. The rocks over which the water there flows are seen to be deeply grooved and worn, every exposed ledge and point having a smoothed and polished look. We can mark how the stone has split up along the natural lines of joint, and how large angular blocks are thus loosened and carried down the stream. In not a few places, too, we may notice cylindrical cavities, called *potholes*, in the bottom of each of which lie a few well-rounded and worn pebbles and boulders. These cavities are due to the circular movements of loose

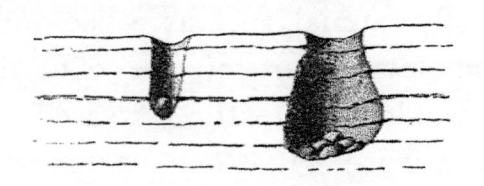

FIG. II.—POT-HOLES EXCAVATED BY RUNNING WATER.

stones that have been caught in eddies, and have been kept whirling there till by their friction they have gradually worked their way downward into the solid rock.

In time, as these cavities are widened and approach each other, their separating walls give way, the hollows unite, and the channel of the stream is thus materially deepened. In dry summer weather, when these *pot-holes* are perchance mere quiet pools of water, their well-worn sides, and the abraded look of the rocks over which the water flows, point in the most impressive way to the progressive erosion of the watercourse. But if we visit the same scene when the stream in flood comes roaring down the rocky gorge with a cumbrous burthen of mud, gravel and stones, the reality of the process is vividly borne in upon the mind. While the boulders are heard striking against each other and upon the rocks of the watercourse as the torrent sweeps them onward, it is not difficult to understand how, under this powerful rasping and grinding, the bottom and sides of the channel should be worn down, and how, by the long continuance of this action, and the other agencies of atmospheric waste, a narrow deep ravine a mile or more in length should be excavated without the aid of any subterranean convulsion.

Having thus loitered for a little by the waterfall and the ravine, let us hasten onward down the valley, and mark how the stream, gathering bulk as it descends, changes character with the variations of its banks. Emerging at last from the gorge, it enters an open part of the valley, and winds in wide, graceful curves from side to side of a flat alluvial meadow. The surface of this plain is raised only a few feet above the average level of the stream, and during floods is often in large part submerged. The river banks are thus low, and as they consist only of soft loam and sand, are in many places undermined and removed by the current. It is observable, however, that where at some bend of the channel the river cuts into the bank on the one side there is usually an accumulation of shingle and sand on the side opposite, so that what is removed from one place is in great measure heaped up at another. These meadow-flats along the margin of streams bear in Scotland the name of "haughs." They have been formed out of the detritus brought down by the streams from the higher parts of the valleys. There are often traces of similar alluvial terraces at higher levels on either side of a valley. Three or four may sometimes be counted rising above each other. They mark each a level which the river once occupied before cutting its way down through its own alluvium to its present channel. Nor is the present level more likely to remain stationary than those which were once in use. The same ceaseless wear and tear is still going on, and the ultimate result must be the same. Even now, indeed, every heavy flood effects an appreciable change both on the bed of the channel and on its sides. Yonder, for example, is a part of the meadow-land where the proprietor with great labour some years ago straightened the river channel, and saved the stream a deal of toilsome turning and winding on its way down the valley. All went well for a time, but a system of vigilant repair was not maintained, and thus the stream, refusing to loose its liberty and flow like a canal, began to bite a bit out of the embankment here and another there; during a heavy storm

> ". . . this river comes me cranking in,
> And cuts me from the best of all my land
> A huge half-moon, a monstrous cantle out;"

and now in the end it has gone back with evident delight to its old twistings and windings among the meadows.

Descending still further, we find the river enter a gorge cut out of boulder-clay, the sides low at first, but rising by degrees to a height of fifty or sixty feet above the channel. The proofs of rapid waste are here far more striking than in the ravine in solid rock just described. Huge masses of clay loosened by recent rains have slipped down the steep bank and project into the current, where they are almost visibly mouldering away, and their detritus is swept down the ravine. Similar land-slips of many different ages, some so old that the fallen masses have put on anew a protecting coat of vegetation, meet the eye on every side. Nor can it be doubted that here, as in the gorge already described, it is the stream itself that has made the excavation. Frost, rains, and springs, loosen the steep banks, and the fallen ruin is worn away by the running water. Thus the ravine is widened and deepened, and though the process may be more rapid here than among hard rocks, yet it is essentially the same in either case. Beyond these ravines the river winds for miles through a fertile champaign country, now stealing quietly among meadows and cornfields, now murmuring through shady woodlands, augmented by many a tributary brook, and bearing its growing width of water through parish and county, until at last, amid sandbars and mudflats, the traces of its own spoils, it enters the sea.[1]

In thus watching the progress of a river from its source to the sea-shore, we learn that the tendency of river-action is to cut out a long, winding channel, narrow or deep in proportion to the nature of the rock, the slope of the ground, the volume of water, and other causes. If a stream acted merely by itself, without the aid of any atmospheric waste, it would cut a series of perpendicular clefts or chasms. This is, perhaps, the reason why in those countries where the rain-fall is comparatively small, or where frosts

[1] The reader who wishes to learn what devastation may be worked by flooded rivers, should read Sir Thomas Dick Lauder's graphic narrative of the Morayshire floods.

are either trifling or unknown, the river-ravines are so profound. Thus the Zambesi in plunging over the precipice at the Victoria Falls enters a gorge 100 feet deep, and only 80 feet broad, which runs in a zig-zag course for many miles. The river seems to have cut its way backward through this winding ravine until, owing to some subterranean movement, effecting a change of level, or to some other cause which would probably be detected by a geologist on the spot, the body of water in place of entering at the top of the ravine has been emptied over one of its sides.[1] By far the most astounding proofs of the power of river-action yet described are those given in the report of the United States Exploring Expedition to the Colorado River of the West. A vast plateau has there been cut by the streams into a net-work of profound chasms often passable neither to man nor beast, but only to the fowls of the air. The scale of these excavations, or cañons, may be estimated from the fact that the Colorado River has cut a channel for itself through the table-land 300 miles long, and with walls rising, often perpendicularly, to a height of from 3,000 to 6,000 feet above the water, which laves their

[1] There is an excellent model of the Victoria Falls in the library of the Royal Geographical Society, which was pointed out to me by Sir Roderick Murchison and Mr. Bates. In looking at it, I was much struck with the resemblance of the so-called "gigantic fissure" to a ravine cut by the action of a stream where springs, rains, and frosts have played only a subordinate part.

D

base. Dr. Newberry, the geologist of the Expedition, states that this network of gorges and chasms is not traceable to "volcanic force, that embodiment of resistless power, that sword that cuts so many geological knots," but "belongs to a vast system of erosion, and is wholly due to the action of water."[1] Where a river not only cuts its way down through solid rock but receives marked aid from springs, rains, and frosts, its channel instead of remaining a deep trench, is widened into a glen or valley by the crumbling away of the sides of the watercourse. Much will, of course, depend upon the nature of the rock, the angle of descent, and, perhaps, on other causes not apparent to us. Where the power of the river is more potent, there will be a tendency to the formation of ravines, and where the influence of atmospheric waste preponderates wider and less precipitous valleys will be the result. I do not mean to assert that a river cannot form a valley, but with reference to the glens and valleys with which I am acquainted, it seems to me that the main office of the rivers, after the hollowing out of their earliest channels, has been to deepen their beds and to carry away the waste of the rocks washed down from either side by rain and springs.

Springs. In the preceding narrative allusion has been made to the co-operative action of springs and frosts. That portion of a rain-shower which does not

[1] Report, Part iii. p. 45.

flow off into the streams, sinks into the earth, filters through the rocks, and collects into underground channels, in which the water again rises to the surface. This subterranean journey is always attended with a waste of the rocks through which the water flows. Part of their substance is disintegrated, or dissolved and carried up to the surface, either in mechanical suspension or in chemical solution. The amount of mineral matter thus removed is very variable. Sometimes it is so abundant as to be deposited when the water, on reaching the air, begins to evaporate. Hence, the ochry scum that gathers upon the stones and grass washed by the outflow of a chalybeate spring, and the white, stony crust that envelopes whatever comes in the way of water that issues out of calcareous rocks. But no spring, even the purest, is wholly destitute of an admixture of such solid ingredients, and thus, the yearly amount of underground waste, due to this cause, cannot but be great. It is the surface-effects of springs, however, that chiefly claim notice here. These are shown in the loosening of masses of rock and earth, from the sides of cliffs and steep slopes. Water filtering through the joints and fissures of the rocks, and always dissolving and carrying away with it part of the sides of the channel through which it flows, loosens the cohesion of a steep bank, and allows portions of different sizes to become detached, and roll down into the watercourse, or valley, below. This kind of waste is well exhibited where a stream flows between banks of boulder-clay, or other loose drift. If we visit such a ravine after a season of rainy weather, we find that the slopes show, in some parts, large faces of freshly-bared clay, or earth, which mark where masses of the bank have been loosened, and launched towards the stream. Large semicircular indentations are thus scooped out along the slope, and by the continuance of this process the ravine grows in width. For the masses detached by the landslips do not permanently protect the slopes behind them; they are attacked by the streams, and in the end removed, so as to leave fresh surfaces exposed to denudation.

Among solid rocks the same process, though commonly at a less rapid rate, may everywhere be seen. There is no long escarpment, or cliff, of such material, which is not traversed with joints, faults, or other divisional planes, that serve as subterranean channels for the water. And as years roll on and the cohesion of the walls of these planes gets loosened, blocks, or large portions of the cliff are, from time to time, detached, and sent to increase the mass of ruin that strews the slope, or lies piled at its foot. The geological structure of the cliff may help to quicken the waste. If, for example, a bed of soft shale should run along the base of an abrupt face of harder rock, the more rapid decay of the soft foundation will cause a proportionately augmented disintegration of the cliff above. A rock full of joints and fractures will usually tend to wear away more quickly than one of equal hardness, but not so jointed, and one which is markedly soluble, as a limestone, generally yields faster to the attacks of time than a hard siliceous stone, such as quartz-rock.

Frost. In close connexion with the disintegrating effects of springs on cliffs, and steep slopes of rock, comes the influence of frost. When the water, which is trickling between the joints of a cliff, is frozen, it expands, and in so doing exerts a vast disparting force on the rocks within which it is confined. On the thawing of the ice the rocks which have been thus separated do not return to their former position; the severance remains until it is increased by another severe frost and thaw. Winter after winter, as the loosened masses are pushed further from each other, the size of the wedge of ice increases, and at last their support gives way, and they fall with a crash from the face of the cliff, leaving a raw scar to mark whence they have come. The effects of frost in pulverizing the soil are a familiar illustration of this disintegrating power. Every one knows that when thaw comes after a long black frost, the country roads are often in as bad condition as after long rain or snow, the reason being that the frost, having separated the particles

also that while the crust is frozen into a cake, the
water, constantly rising from below accumulates

of the clay, has allowed the thawed moisture to mix thoroughly with them.

It will thus be seen that water filtering through the rocks, and frost solidifying it there, are powerful auxiliaries to a river in the working out of its ravines. To these combined influences, also, the present outlines, if not the first origin, of most inland cliffs should probably be ascribed. No one can look at such precipitous faces of rock, with the usual scattered fragments at their base, without being convinced that in what way soever they may have been formed at first, they are slowly yielding to the powers of atmospheric waste, and that they are thus by degrees creeping back, just as the waterfall of a ravine is retreating. The line occupied by the cliff of an escarpment now stands far removed from where it was when the cliff began to be formed. I am strongly inclined with my colleague, Professor Ramsay, to view this recession as mainly, if not wholly, due to the slow wasting powers of the atmosphere. In such an explanation no hypothetical agent is called into play, and it can hardly be denied that if the sources of decay which we see, in the present day, at work upon these cliffs were allowed a long enough time, they could do all that is needed of them.

CHAPTER III.

ACTION OF THE SEA AND OF WIND.

IN contemplating the gradual waste to which the surface of the earth is everywhere subjected, the observer is soon struck with the signal proofs of decay furnished by that outer border of the land which is washed by the sea. The abrupt cliffs that shoot up from high-water mark, the skerries that rise among the breakers a little way from the shore, and the sunken reefs that lie still further out to sea—all tell of the removal of masses of solid rock. A little reflection leads us to perceive that the abrading power of the sea must in all likelihood be confined to that upper part of the water which is affected by winds and tides, and that in the deeper abysses there is probably no actual erosion of hard rock, though the currents there may be capable of carrying along fine ooze and silt.

The waste that takes place along the line where land and sea meet has a twofold character. In the first place there is a direct abrasion by the sea itself when it throws its breakers, often loaded with coarse shingle, against the coast, and wears down even the hardest rocks. In the second place, cliffs and precipitous banks overlooking the waves, are subject to that never-ceasing atmospheric waste described in the last chapter; and the sea, in many places, does little more than remove from tide-mark the débris that has been loosened from the cliffs by springs and frosts. Thus sea-cliffs, like the walls of river-gorges, receive no permanent protection from the accumulation of their own ruins in front of them. In looking at the results of the wear and tear along a coast-line, we are apt to assign too much importance to the mere mechanical impetus of the breakers, and too little to the less obtrusive but more constant influences of rains, springs, and frosts. It may be impossible to give to each agency its due share in the wasting of a coast-line, but it should not be forgotten that in what is usually called *marine denudation*, the atmospheric influences play a great part.

No one can watch the progress of a storm on an exposed rocky coast without being strongly impressed with the powerful effects of breakers in wearing away the margin of the land. A wave which can deal a blow equal to a pressure of 6,083 pounds on the square foot (and such is the ascertained impetus of storm waves among the outer Hebrides) is no feeble instrument of abrasion. Yet such a wave can of itself have little or no power to grind down the surface of the rocks on which it beats, for that surface, even after a storm, is found to be just as plentifully coated with living barnacles as before. If the friction of the water could rub down the stone these cirripedes would be removed first. It is only where its enormous weight and impetus can break off a loosened mass of rock that a wave may be said to act by its own sheer force. In the great majority of cases, however, breaker-action eats into a coast-line by battering down the rocks with their own débris. A wave that lifts up, and sweeps forward gravel, boulders, and even large blocks of stone, is a far more formidable instrument of destruction than even a large wave which is not armed with the same weapons. The stones that are thus swung on by the tempest fall with prodigious force against the rocks of the shore; brought back again by the recoil of the wave they are caught up by its successor and again hurled forwards upon the rocks. And thus by what has been aptly termed a kind of sea-artillery, even the hardest rocks of an iron-bound shore are worn away.

To see these features in perfection, the observer should betake himself to some rocky shore on which falls the full roll of the Atlantic. He will then find, if the coast be a precipitous one, that the rocks above the reach of the waves are rough and rugged, showing everywhere traces of that subaërial waste which, acting along their natural joints, has slowly shattered the crags and sent down huge blocks to the beach below. There the fallen ruin coming within reach of the waves, is turned into a further means of augmenting the destruction of the cliffs. Ground down by the waves into well-worn boulders it is driven up against the cliffs, which along their base are smoothed and polished like the shingle. The line between the rough surface overhead, marking the progress of the atmospheric waste, and the well-worn zone of the beach, pointing to the work of the sea, is often singularly sharp. But the base of the cliff is not merely polished by the friction of the boulders: it is in many places hollowed out into overhanging recesses, clefts, and caves. At the further end of such excavations we often come upon the rounded boulders—wet with the last tide—that are carrying on the work of demolition, and we may then see how the waves, breaking against the base of the cliffs, and rushing furiously into every cave and cranny to which they can reach, hurl the stones backward and forward against the solid rock. Between the upper limit of the tide and low-water mark crags and skerries may be noted in every stage of decay. Here we mark an outjutting mass of the cliff which has been separated, with the breakers meeting and bursting into foam in the narrow passage. Yonder a mass, once evidently connected with the main cliff in the same way, has been sundered by the roof of the tunnel falling in, and it now stands up as a tall massive outwork of the line of rampart behind. Lower on the narrow beach, worn, tangled-covered bosses of rock rise out of the shingle and boulders, and run out to sea in low reefs that are usually fringed with foam even in the calmest summer day. Here and there, these sunken skerries shoot up into islets haunted by seal and wild-fowl. Whether we take boat and row quietly among these reefs and islands, or pick our way toilsomely over the rocky beach, or ramble along the summit of the crags overhead, we are everywhere met with proofs of unceasing destruction. We see that the cliffs must once have stretched seaward, at least as far as yonder seastack, fully a furlong from their present limit, and how much further no man can tell. It is impressively taught that the selvage of land which has been cut off has been carried away by the sea. The whole process in all its stages is before our eyes. We note the weather wasting the cliffs above, and the sea battering them below. And it is impossible to doubt, that if in a comparatively short geological period a strip of land, say a furlong broad, has been in this way planed down, there is here revealed to us a power of waste, the effects of which have perhaps no limit short of the total demolition of the dry land.

In looking more narrowly at the progress of this abrasion, we find it dependent not merely on the prevalent winds and the consequent *fetch* of the breakers, but in large measure upon the varying geological structure of the coast-line. The waste of the eastern shores of the British Isles is thus much more rapid than that of the western, and the reason is, that though the waves of the German Ocean are not so powerful as those of the Atlantic, yet the rocks forming the coast-line on that side are, as a whole, more easily worn away than those on the west side. If the soft sandstones and shales, clays and sands of the eastern sea-board were open to the full fury of the western ocean, there would be a sad yearly tale of loss and ruin. Perhaps, it may be objected that the western coast is far more indented with inlets and fiords than the eastern, and therefore shows more strikingly the wasteful powers of the sea. But these indentations, as we shall afterwards notice, are not the work of the sea; they are, in truth, submerged land-valleys, and point to the prolonged action of subaërial waste on a wide terrestrial surface, of which our present islands are only a small part.

It would be beyond the scope of this little volume to enter into the details of the relation between geological structure and the action of the waves. But it may in part serve the purpose, and show at the same time how real is this source of decay, if we inquire what has been the nature and amount of the loss of land which is known to have occurred along our shores in recent times. It would be interesting if we could trace the gradual retreat of the coast-line since man became an inhabitant of this country, or even since the time to which the earliest historical notices of Britain refer. No written records of such changes, however, go further back than, at the most, three or four hundred years. There are, indeed, traditions of land having once existed, where for many a century have rolled the waves of the salt sea; just as in Cornwall there still survives the memory of a district, called the Lionnesse, now covered by the Atlantic, but which in the days of the Knights of the Round Table is said to have been rich and fertile. But such traditions are too vague to be, at least in the meantime, of any geological service. It is with the time of written history therefore that we must deal:—in short with the changes that have taken place along the coast-line within the last few hundred years. The period during which observations have been recorded is thus but of short duration, yet it furnishes us with some instructive lessons as to the progress of the marine erosion, and enables us in some measure to see how the decay of the coast-line has gone on in time past. Let the reader imagine himself coasting northwards along the eastern sea-margin of Scotland, and while the breeze drives us merrily onward, it may be a pleasant amusement to listen to some jottings of the wild havoc that has been brought on the shores, by the same sea whose waves are now leaping and laughing around us.

We set sail from the mouth of the Tweed, and skirt the abrupt rocky coast which forms the seaboard of Berwickshire. The cliffs for many miles are steep or vertical, rising, near St. Abb's Head, to a height of five hundred feet above the waves, and here and there interrupted by narrow bays and coves, which have in several instances been selected as the sites of fishing villages and hamlets. We see from the wasted and worn look of these cliffs what a sore battle they have had to fight with the ocean. Craggy rocks, isolated stacks and sunken skerries, that once formed part of the line of cliff, are now enveloped by the restless waves. Long twilight caves, haunted by otters and sea-mews, and flocks of rock-pigeons, have been hollowed out of the flat carboniferous sandstones and the contorted Silurian greywacké, and are daily filled by the tides; and then in storms, the whole of these vast precipices, from base to summit, is buried in foam—the pebbles and boulders, even on the sheltered beaches, are rolled back by the recoil of the breakers, and hurled forward again, with almost the force and noise of heavy cannon. But a line of abrupt rock presents such formidable obstacles to the advance of the sea, that the rate of waste is extremely slow. We see everywhere, indeed, from the geological structure of the coast, that the loss of land must have been great, and that it is still in progress. But there does not appear to be any record to show how much has taken place within the times of human history. Passing onward, therefore, along this coast, with its green bays and dark gloomy cliffs, we round the headland of St. Abb's, and observe that it stands there, at once a bulwark against the waves and a mark of their advance; for, being a mass of hard porphyry, it has been able in some measure to withstand the assaults of the ocean which has worn away the greywacké and shales around. Sweeping across the Bay of Dunglass we pass the sandstone cliffs of Cove (where the old fishermen point to great inroads by the sea during their lifetime), and then the shores of Skateraw, where, in the early part of this century, stood the ruins of an old chapel, which were swept away many years ago, the tides now ebbing and flowing over their site.[1] Further west stands the Castle of Dunbar at the entrance into the Frith of Forth. There the proofs of degradation and decay come before us with a melancholy reality. The old castle, once so formidable a stronghold, is almost gone—two tall fragments of wall, and some pieces of masonry at a lower level being all that is left. It is not merely that the rains and frosts of many a dreary winter have broken down the ramparts, nor even that the hand of man, more wanton and unmerciful in its destruction than the hand of time, has quarried the stones and blasted the rocks in the excavation of the harbour. The sea has been ceaselessly at work wearing away the islets and the cliff on which the ruin is perched, and the time will come, at no very distant date, when the *Dun* or hill from which the castle takes its name, will be swept away, and its site be marked only by a chain of rocky skerries. A little to the west of the castle, a huge mass of the sandstone cliffs, undermined by the sea, fell during the night a year or two ago. The scar is yet fresh, and though

---

[1] Popular Philosophy, or the Book of Nature, laid open upon Christian principles. Dunbar, 1826, vol. ii. p. 160.

a pile of ruin still lies at the foot of the precipice, it is being broken up and carried away by the waves.

It might have been supposed that the comparatively sheltered estuary of the Firth would be free from any marked abrasion by the sea, yet even as far up as Granton, near Edinburgh, during a fierce gale from the north-east, stones weighing a ton or more have been known to be torn out of a wall and rolled to a distance of thirty feet. Hence, within the last few generations, the sea has made encroachments, sometimes to a considerable extent, along the whole coast of the Firth, even as far up as Stirling. Tracing the southern shores in a westerly direction from Dunbar, we find that the low sandy tracts at the mouth of the Tyne, and again from North Berwick to Aberlady, have suffered loss in several places. Further on, near Musselburgh, there was a tract of land on which the Dukes of Albany and York used to play at golf in former days, but which is now almost entirely swept away. The coast of Edinburghshire has in like manner lost many acres of land. Maitland, for instance, in his History of Edinburgh, describes the ravages of the sea between Musselburgh and Leith which had occasioned the "public road to be removed further into the country and the land being now violently assaulted by the sea on the eastern and northern sides, all must give

[1] Stevenson, Trans. Royal Soc. Edin. xvi. 27.

way to its rage, and the links of South Leith probably in less than half a century will be swallowed up."[1] The road alluded to has had to be removed again and again since this passage was written. Mr. Stevenson[2] remarked in 1816 that even the new baths, erected but a few years before at a considerable distance from the high-water mark, had then barely the breadth of the highway between them and the sea, which had overthrown the bulwark or fence in front of those buildings and was then acting on the road itself. Maitland speaks also of a large tract of land on both sides of the port of Leith, which has likewise disappeared. Nor are the inroads of the sea less marked as we continue our westward progress. The old links of Newhaven have disappeared. If the calculations of Maitland may be believed,[3] three-fourths of that flat sandy tract were swallowed up in the twenty-two years preceding 1595. Even in the early part of the present century, it was in the recollection of some old fishermen then alive, that there stretched along the shore in front of the grounds of Anchorfield, an extensive piece of links on which they used to dry their nets, but which had then been entirely washed away. The direct road

---

[1] History of Edinburgh, p. 499.
[2] In an excellent paper in the second volume of the Werneran Society's Memoirs, to which I have been greatly indebted in collecting the statistics given above.
[3] History of Edinburgh, p. 500.

between Leith and Newhaven used to pass along the shore to the north of Leith Fort, but it has long been demolished, and for at least fifty years the road has been carried inland by a circuitous route.[1] The waste still goes on, though checked in some degree by the numerous bulwarks and piers which have been erected along the coast. The waves impinge at high tides upon a low cliff of the stiff blue till or boulder-clay which readily yields to the combined influences of the weather. Hence large slices of the coast-line are from time to time precipitated to the beach. A footpath runs along the top of the bank overhanging the high-water mark, and portions of it are constantly removed on the landslips of clay. By this means, as the ground slopes upwards from the sea, the cliff is always becoming higher with every successive excavation of its sea-front. The risk to foot-passengers is thus great; so many accidents, indeed, have occurred here that the locality is known in the neighbourhood as the *Man-Trap*.[2]

Higher up the Firth of Forth, at the Bay of Barn-

[1] Stevenson, Mem. Wer. Soc. vol. ii.
[2] Since the above was written the man-trap has ceased to be; not, however, by the destructive force of the waves, but under the combined operations of mattock, wheelbarrow, and waggon. A branch railway has been brought along the coast-line, and the accumulated rubbish from some long cuttings through boulder clay has been shot over the sea cliff, completely covering it up, and thus carrying the land out to sea again. A large piece of ground has thus been reclaimed. It is to be protected by a bulwark.

bougle, a lawn of considerable extent, once intervening between the old castle and the sea, has been demolished. Even in the upper reaches of the estuary, above the narrow strait at the Ferries, the waves have removed a considerable tract of land, which once intervened between the sea and the present road leading westward from Queensferry. Similar effects have likewise been produced on the northern shores of the Firth, at Culross and eastwards by St. David's, Burntisland,[1] Kirkaldy, and Dysart. The seaports along this coast have all suffered more or less from encroachments of the sea—roads, fences, gardens, fields, piers, and even dwelling-houses having been from time to time carried away. In the parish of Crail some slender remains of a priory existed down to the year 1803. These, along with the old gardens and fences, are now wholly removed; but the adjoining grounds still retain the name of the croftlands of the priory. At St. Andrew's, Cardinal Beaton's Castle is said to have been originally some distance from the sea; but it now almost overhangs the beach, and must ere long fall a prey to the waves.[2]

[1] "At the east end of the town of Burntisland the sea comes now far in upon the land; some persons in the town, who died not long since, did remember the grassy links reach to the Black Craigs, near a mile into the sea now."—Sibbald's *Fife and Kinross*, p. 152. The waste still continues, in spite of the strong railway embankment, much damage having been done by the storms of last winter (1864-5).

[2] Stevenson, *Ibid.* "The learned Mr. George Martine (*Reliquiæ Sancti Andreæ*, chap. ii. p. 3) relates it as a tradition received that the

Passing northwards along the eastern coast of Scotland we find that the sea has encroached to a marked extent on the sands of Barry, on the northern side of the Firth of Tay. The lighthouses which were formerly erected at the southern extremity of Button Ness have been from time to time removed about a mile and a quarter further northward, on account of the shifting and wasting of these sandy shores. The spot on which the outer lighthouse stood early in the seventeenth century was found to be in 1816 two or three fathoms under water and at least three-quarters of a mile within flood mark.

If the waves can bring about such important changes, even when rolling into more or less sheltered estuaries, we may expect that their power will be found still greater where, without any bounding land to curb their fury, they can rush in from open sea, and fall with unbroken violence upon an exposed coast-line. That this is the case

ancient Culdees, Regulus and his companions, had a cell dedicated to the Blessed Virgin about a bow flight to the east of the shoar of St. Andrews, a little without the end of the peer (now in the sea), upon a rock called at this day Our Lady's Craig: the rock is well known, and seen every day at low water. The Culdees thereafter, upon the sea's encroaching, built another house where the house of the Kirkheugh now stands called Sancta Maria de rupe, with St. Rule's Chapel, and says in his time there lived people in St. Andrew's who remembered to have seen men play at bowls upon the east and north sides of the castle of St. Andrew's, which now the sea covereth at every tide."—Sibbald's Fife and Kinross, p. 152.

with the German Ocean is shown by the form of the coast-line, the known effects of storms, and by actual experiment of the power of the breakers. The force with which the waves of this ocean fall on objects exposed to their fury has been measured with great care at the Bell-Rock Lighthouse. This massive structure, rising 112 feet above the sea-level, "is literally buried in foam and spray to the very top during ground swells when there is no wind." Experiments were made there from the middle of September, 1844, to the end of March, 1845, and the greatest recorded pressure was 3013 pounds on the square foot. Mr. Stevenson, however, under whose direction the observations were conducted, informs us that on the 27th November, 1824, the spray rose 117 feet above the foundations, being equivalent to a pressure of nearly three tons on the square foot.[1] Such enormous force cannot but produce marked effects on all rocks exposed to its fury. In May, 1807, during the building of the lighthouse six large blocks of stone, which had been landed on the reef, were removed by the force of the sea and thrown over a rising ledge to the distance of twelve or fifteen paces, and an anchor, weighing about 22 cwt. was thrown up upon the rock.[2]

[1] Trans. Roy. Soc. Edin. xvi. 28.
[2] Account of Erection of Bell-Rock Lighthouse, p. 163.

Coast of Forfarshire.

This power of transport affecting parts of the surrounding sea-bed during severe gales has been frequently observed here. Stones measuring upwards of 30 cubic feet, or more than two tons in weight, have often been cast upon the reef from deep water.[1] These large boulders are so familiar to the light-keepers at this station as to be by them termed *travellers*.[2]

With breakers of such prodigious force beating winter after winter on its sands and rocks, the eastern coast of Scotland suffers sorely as the years roll on. Nowhere does it exhibit more striking proofs of the unavailing resistance which it offers to the ocean than along the borders of Forfarshire. There we see some of the wildest scenery on this side of our island: huge beetling cliffs of red sandstone washed and worn, crags of hard trap rocks that seem ready to topple into the surf, creeks in which the gurgling tides are for ever rolling to and fro, caves sometimes out of reach of the waves, and then coated with mosses and ferns, sometimes at a low level and filled well nigh to the brim when the tide runs at its full, while the space between tide-marks is a chaos of craggy rocks and skerries and huge boulders torn from the cliffs overhead. And what has caused

---

[1] The sea at a distance of 100 yards all round the sunken reef of the Bell Rock has a depth of two or three fathoms at low water.
[2] Edin. Phil. Journ. iii. 54.

this wild ruin? Not any cataclysm or convulsion of nature, no earthquake, no outbreak of volcanic fire. It has been done partly by the rains and frosts of ages, and partly by yonder waves that seem to curl so peacefully along the distant strip of sandy shore and break into little eddies of foam around the nearer rocks; but which, when the north-east gales sweep across the sea, batter against the cliffs with the noise of thunder, and cover them with spray even to the summit. The Forfarshire coast-line is, for the most part, formed of such wall-like cliffs of red sandstone. But here and there, in creeks and bays, there are sandy flats—the records of an older sea margin yet to be described. It is upon these softer parts that the breakers have made most rapid inroads. Thus, in the thirty years that preceded 1816, the road trustees were under the necessity of twice removing inland the road-way that skirts the shore westward from Arbroath, and in that year it had again become imperative to make another removal. The loss of land at one point, a short way south-west from the town, has been thirty yards since 1805, while at another spot still nearer the town, it has reached as much as sixty yards within the same period—that is, at the rate of fully a yard every year. About the year 1780 a house existed at the latter locality, of which there are now no remains, its place being covered by the tides. At Arbroath itself a house which stood next to the sea was a few years ago washed down, and strong bulwarks are necessary to retard further encroachments.[1] But these prove to be ineffectual barriers, for every severe gale damages them, and the sea is sensibly gaining ground.

The coast as we proceed northwards continues to furnish additional instances of the destructive effects of the sea within the historical period. "On the Kincardineshire coast," says Sir Charles Lyell,[2] "an illustration was afforded at the close of the last century of the effect of promontories in protecting a line of low shore. The village of Mathers, two miles south of Johnshaven, was built on an ancient shingle beach, protected by a projecting ledge of limestone rock. This was quarried for lime to such an extent that the sea broke through, and in 1795 carried away the whole village in one night, and penetrated 150 yards inland, where it has maintained its ground ever since, the new village having been built further inland on the new shore." In order to check the further ravages of the waves, a stone bulwark was erected, which is still kept up for the protection of the houses which stand nearest the beach.[3]

[1] Stevenson, *Op. cit.*
[2] Principles of Geology, ninth Edition, p. 302.
[3] New Stat. Acc. Kincardine, p. 275.

The shores of the Moray Firth afford several instances of the destruction by the sea of ancient buildings. Thus, at the old town of Burghead, on the eastern headland of Burghead Bay, a fort, said to be of Danish origin, was built upon a sandstone cliff, between which and the sea, according to tradition, there once lay a very considerable tract of land, but the ruin now actually overhangs the waves.

A few miles westward on the same coast stands the town of Findhorn, which has been the scene of extensive devastations. The shore is low and sandy, and is liable to change its outline, owing to the constant drifting of the sandhills. Between these ridges of sand and the sea margin, there runs along the parish of Kinloss, west of Findhorn, a band of coarse gravelly shingle, which acts to some extent as a bulwark against the waves. But that it has proved an ineffectual barrier is shown by the fact that the present village of Findhorn is the third that has borne the name. The first stood about a mile west of the bar, the point at which the river originally entered the Firth, before the eastward progress of the moving sand drove it into the channel it now occupies. The second village was planted a little to the north of the present one, but it, too, has been swept away. Nor does it appear that the existing town is free from the risk of being overtaken, partially at least, by a similar catastrophe. "The little space that intervenes between the tide-mark and the north end," says the reverend statist of the parish, "is a broken bank of sand that drifts dreadfully with every hurricane, covering the streets and gardens to the depth of sometimes eight or ten feet, and this constitutes but a feeble bulwark against the tremeudous surf that beats with a north-easterly swell; so that if means be not taken to give it a solid surface, either by laying it over with turf or planting it with bent, there is reason to apprehend that it will by and by be blown away altogether, leaving Findhorn that now is to share, at some future period, the fate of its predecessors."[1] Even into the recesses of the Moray Firth the ocean carries with it its resistless power of demolition. Thus, encroachments that had been made on the coast round Fort George early in this century, were such as to raise fears for the safety of the fortress. Some of the projecting bastions, previously at a distance from the sea, were then in danger of being undermined by the water; and it was found necessary to break the force of the waves by erecting a sort of *chevaux de frise* in front of the walls. On the north shore of the Beauly Firth, a number of sepulchral cairns have been engulphed by the sea. One of these stands fully 400 yards within tide-mark, and it has been calculated that an area

[1] New Stat. Acc. Scotland, Nairn, p. 203.

of not less than ten square miles, now flooded by the advancing tide, has been the site of the dwellings of the ancient cairn-builders.[1]

The long sheltered estuary of the Cromarty Firth, so thoroughly land-locked that it communicates with the open sea only through a narrow channel between the headlands of the two Sutors, might be supposed to be free from any risk of attack by the waves. Yet even there the same tale of waste is told. It is said by Hugh Miller that the tide now flows twice every twenty-four hours over the spot where a hundred years ago there stood a pedlar's shop.

Along the cliffs of the Caithness shores every winter gives fresh proof of the immense destructive power of the breakers of the North Sea. At some places, in particular to the south of the town of Wick, the waves have quarried out masses of Old Red Sandstone and piled them up in huge heaps on the top of the cliff, sixty or a hundred feet above high-water mark. Some of the blocks of stone which have been moved from their original position at the base or on the ledges of the cliffs, are of great size. My friend Mr. C. W. Peach has been so kind as to send me some notes regarding them. "The largest disturbed mass," he says, "contains more than 500 tons, and is known as Charlie's Stone.

---

[1] Wilson's Prehistoric Scotland, p. 63.

Others varying in bulk from 100 to 5 tons or less lie by hundreds piled up in all positions in high and long ridges, which, before the march of improvement began in the district, extended far into the field above the cliff.[1] Near the old limekiln, South Head, similar large blocks of sandstone have been moved by the gales of the last three years. The great storm of December, 1862, in particular distinguished itself by the havoc which it wrought along these shores. It swept the sea over the north end of the island of Stroma, which lies in the Pentland Firth,[2] and redistributed the ruin-heaps there. The waves ran bodily up and over the vertical cliffs on the west side, 200 feet in height, lodging portions of the wrecked boats, stones, seaweeds, &c. on the top.

[1] This mass of ruin was noticed by Hugh Miller, who suggested that it might have been produced by the stranding of icebergs. Mr. Peach, however, remarks that there is every reason to believe that it is the work of the sea, and even now an occasional stone is added to the pile—the last having been a block at least half a ton in weight, which about two years ago was torn up from its position fifty feet above the sea-level.

[2] The spring tides of the Pentland Firth are said to run at a rate of nine nautical miles in the hour, while the rate of the neap tides is three miles. These are probably by much the most rapid marine currents round the British Islands. Yet that they do not of themselves produce any appreciable abrasion of the coast-line is shown by the coating of barnacles and seaweed on the rocks even at low water. As currents, their power by mere friction is probably *nil;* but when they are aided by powerful winds they lend a prodigiously accelerated impetus to the ordinary wind-waves. Hence the incredible force of the breakers in these northern seas.

They rushed in torrents across the island, tearing up the ground and rocks in their course towards the old mill at Nethertown on the opposite side. This mill had often before been worked by water collected from spray thrown over these cliffs, but never had such a supply been furnished as by this gale. One curious phenomenon was noticed at the south end of Stroma; the sea there came in such a body between the island and the Caithness coast, that at intervals it rose up like a wall, as if the passage was too narrow for the mass of water which, forced onwards from the Atlantic between Holburn Head on the Caithness shore and the Old Man of Hoy on the Orkney side, passed bodily over the cliffs of Stroma. The effects of this terrific gale will long be remembered. Some time after its occurrence I was on Stroma and along the Pentland Firth side, and was deeply struck with the ruin spread around. The huge masses that had been moved exceeded all I had ever seen before. With this evidence, added to a long experience of storms, I am compelled to believe that the ruin of cliffs and the heaping-up of torn rock-masses have been effected by the sea when agitated by storms, and not by icebergs."

Crossing the wild tides of the Pentland Firth, we find ourselves among the rocky fiords and voes of the Orkney and Shetland Islands. There the power of the sea comes before us even more impressively. The intricate indented coast-line, worn into creeks and caves and overhanging cliffs; the crags, and skerries, and sea-stacks, once a part of the solid land, but now isolated among the breakers; the huge piles of fragments that lie on the beach, or have been heaped far up above the tide-mark, tell only too plainly how vain is the resistance even of the hardest rocks to the onward march of the ocean. The rate of waste along some parts of these islands is so rapid as to be distinctly appreciable within a human lifetime. Thus, in the chain of the Orkneys, the Start Point of Sanday was found by Mr. Stevenson in 1816 to be an island every flood tide; yet even within the memory of some old people then alive, it had formed one continuous tract of firm ground. Nay, it appears that during the ten years previous to 1816 the channel had been worn down at least two feet.

Probably no part of the British coast-line affords such striking evidence of the violence of the waves as that which may be seen along the margin of the Shetlands. These islands are exposed to the unbroken fury at once of the German Ocean and of the Atlantic, while the tides and currents of both seas run round them with great rapidity. Hence their seaboard wears in many places an aspect of utter havoc and ruin. Against their eastern side the North Sea expends its full violence, tearing up the rocks from the craggy headlands, and rolling onwards far up into the most sheltered fiords. The island of Whalsey, for instance, lying off the east side of the mainland, about the middle of the Shetland group, is completely sheltered from the gales of the Atlantic. Yet in the Bound Skerry of Whalsey, the breakers of the North Sea have torn up masses of rock sometimes 8½ tons in weight, and have heaped them together at a height of no less than 62 feet above high-water mark. Other blocks, ranging in bulk from 6 to 13½ tons, have been actually quarried out of their place *in situ* at levels of from 70 to 74 feet above the sea. One block of 7$\frac{7}{10}$ tons, lying 20 feet above the water, has been lifted from its bed and borne to a distance of 73 feet from S.S.E. to N.N.W. over abrupt opposing forces of rock as much as seven feet in height.[1] On the west side of the Shetland Islands the fury

[1] See an interesting paper by Mr. Stevenson. Proc. Roy. Soc. Edin. iv. 200. See also his work "On the Design and Construction of Harbours" (1864), pp. 30—38. Mr. Peach, in a paper on the traces of Glacial Drift in the Shetland Islands, read before the British Association in 1864, notices further proof of the power of the breakers among these islands. He states that on the top of the cliffs of the island of Honsay, about 100 feet high, the waves break in stormy weather, tearing up the rock and piling its huge fragments into a semi-circular wall a considerable way back from the edge of the cliff. "Between this wall and the cliff a deep river-like gully is scooped out, down which the water rushes again to the sea, a great distance from the spot whence it was thrown up."

of the Atlantic has produced scenes of devastation which it is hardly possible adequately to describe. In stormy winters, huge blocks of stone are overturned or are removed from their native beds to a distance almost incredible. Dr. Hibbert found that in the winter of 1802 a tabular mass, 8 feet 2 inches in length by 7 ft. in breadth and 5 ft. 1 in. in thickness, was dislodged from its bed and removed to a distance of from 80 to 90 feet. In 1820, he found that the bed from which a block had been carried the preceding winter, measured $17\frac{1}{2}$ ft. by 7 ft. and 2 ft. 8 in. in depth. The removed mass had been borne a distance of 30 feet, when it was shivered into 13 or more fragments, some of which were carried still further, from 30 to 120 feet. A block of 9 ft. 2 in. by $6\frac{1}{2}$ ft. and 4 ft. thick, was hurled up the acclivity to a distance of 150 feet. "Such," he adds, "is the devastation that has taken place amidst this wreck of nature. Close to the Isle of Stenness is the Skerry of Eshaness, formidably rising from the sea, and showing on its westerly side a steep precipice, against which all the force of the Atlantic seems to have been expended: it affords refuge for myriads of kittiwakes, whose shrill cries, mingling with the dashing of the waters, wildly accord with the terrific scene that is presented on every side."[1]

[1] Hibbert's Shetland Islands, p. 527.

The result of this constant lashing of the surge has been to scarp the coasts of the Shetlands into the most rugged and fantastic cliffs, and to pierce them with long twilight caves. Dr. Hibbert describes "a large cavernous aperture, 90 feet wide, which shows the commencement of two contiguous immense perforations, named the Holes of Scranda, where in one of them that runs 250 feet into the land, the sea flows to the utmost extremity. Each has an opening at a distance from the ocean, by which the light of the sun is partially admitted. Further north other ravages of the ocean are displayed. But the most sublime scene is where a mural pile of porphyry, escaping the process of disintegration that is devastating the coast, appears to have been left as a sort of rampart against the inroads of the ocean. The Atlantic, when provoked by wintry gales, batters against it with all the force of real artillery—the waves having, in their repeated assaults, forced for themselves an entrance. This breach, named the Grind of the Navir, is widened every winter by the overwhelming surge, that, finding a passage through it, separates large stones from its side, and forces them to a distance of 180 feet. In two or three spots the fragments which have been detached are brought together in immense heaps, that appear as an accumulation of cubical masses, the product of some quarry."[1] In other places the progress of the ocean has left lonely stacks, or groups of columnar masses at a distance from the cliffs. Such are the rocks to the south of Hillswick Ness, and the strange tower-like pinnacles in the same neighbourhood called the Drenge, or Drongs, which, when seen from a distance, look like a small fleet of vessels with spread sails.

The Hebrides not less than the Shetlands illustrate the power of the ocean in working the degradation of the land. The most careful observations of the force of the breakers in this part of the British seas are those made under the direction of Mr. Stevenson, during the progress of the erection of the lighthouse on Skerryvore—a rock lying to the south-west of the island of Tiree, and exposed to the full fury of the Atlantic, there being no land between this point and the shores of America. The average results of these experiments for five of the summer months during the years 1843 and 1844, give to the breakers a force of 611 lbs. per square foot; and for six of the winter months of the same years 2,086 lbs. per square foot, or thrice as great as in the summer months. The greatest result obtained was during the heavy westerly gale of 29th March, 1845, when a pressure of 6,083 lbs. per square foot was registered. This was a force of little short of three tons on every

[1] Hibbert's Shetland Islands.

square foot of surface. The next in magnitude was a force of 5,323 lbs.[1]

North-west of Skerryvore lies the island of Barra Head, the last of the long chain of the broken and deeply embayed Hebrides. It is recorded that on this island, during a storm in January, 1836, a mass of gneiss containing 504 cubic feet, and estimated to be about 42 tons in weight, was gradually moved five feet from the place where it lay, having been rocked to and fro by the waves, until a piece broke off, which, jamming itself between the block and the rock below, prevented any further movement.[2]

Fortunately the rock which has had to withstand this tremendous battery, is a tough gnarled gneiss. But where the coast is low, and more especially where the hard gneiss passes under a covering of blown sand, the Atlantic breakers have made sad inroads even within the last few generations. "The most destructive process of nature," says the author of the description of the Isle of Harris in the old Statistical Account of Scotland, "is the continual wasting of the land on the western shore by the perpetual drifting of the sand, and the gradual encroachment of the sea. This is evinced by the clearest testimonies. Lands which were ploughed within the remembrance of people yet living, are now no more. Wherever a high sand-bank has

[1] Stevenson, Trans. Roy. Soc. Edin. xvi. 25. [2] *Ibid*, p. 28.

been entirely worn away the soil under it is found to have been either a rich loam or black moss. In many such situations, vestiges of houses, enclosures, churches, and burying grounds appear."[1]

This chain of islands, which, like a great breakwater fronts the western coast of Scotland, has doubtless preserved this side of the country from not a little of the destruction which would otherwise have fallen upon it. The greater hardness of the rocks, as compared with those of the east coast, must also have contributed to retard in some measure the progress of the waves. Nor must we forget that the absence of harbours and maritime villages and towns on the western sea-board has probably deprived us of a record of the waste of these shores within the historical period. Knowing the actual force of the waves, and seeing how much they can effect in a stormy winter, we cannot doubt that during the last few hundred years there must have been more or less loss of land, even along the most iron-bound parts of the coast, although no memorial remains to tell how and when exactly the loss was effected. That there must have been considerable waste along the more exposed shores may be inferred from the fact, that even in the Kyles of Bute, a tract of low land to the north of Kames Bay has been so encroached upon by the tides,

[1] Old Stat. Acc. vol. x. p. 373.

that a road which skirted the beach has been thrice removed further inland within the last thirty or forty years.[1]

Along the shores of the estuary of the Clyde the sea has in some places removed a considerable part of the coast-line even within recent times. Thus to the south of the town of Ayr, a cliff of volcanic ash rises vertically from the beach, bearing on its verge the picturesque ruin of Greenan Castle. The walls overhang the precipice, and the sea is hollowing out the rock below. Yet within the recollection of a venerable lady who died some years ago, there was room for a horse and cart to pass between the castle and the edge of the cliff. During the last hundred years, therefore, a slice of solid rock, perhaps six or eight feet broad, has been cut away from this part of the coast. A short distance further south a spring in the middle of a field, a few feet above high-water mark, was enclosed as a well some sixty or seventy years ago. Since then that part of the field which lay between the well and the sea has been eaten away, and the spring now rises at the edge of the shingle of the beach.[2]

The shores of Loch Ryan, which seem so well

---

[1] My much esteemed friend the Rev. Alexander Macbride, of Ardmory, Bute, pointed this fact out to me.
[2] For these facts I am indebted to Dr. Sloan, of Ayr.

sheltered alike from the Atlantic and the Irish Sea, have suffered considerably within the last two or three generations. Mr. Stevenson found in 1816 that at the town of Stranraer the houses along shore had formerly gardens between them and high water, but that of late years the inhabitants had been under the necessity of erecting bulwarks to secure the walls and approaches to their houses. Further down the loch, at the village of Kirkolm, a neck of land called the Scar Ridge had once extended into the sea about half a mile. Cattle were wont to graze upon it, but it was then nearly washed away, and in high tides it was laid almost wholly under water.[1]

The southern coast-line of Scotland lies open to the full fury of the Irish Sea. When the wind blows strongly from the south-west the rocky precipitous shores of Wigton and Kirkcudbright are white with foam, headland after headland standing out into the breakers that roll eastward far up into the recesses of the Solway Firth. In a series of experiments made during the fine summer of 1842, at the Island of Little Ross, on the coast of Kirkcudbright, it was found that the average force of the waves was about 328 lbs. on the square foot, or rather more than half the average summer force of those at

[1] Stevenson, Mem. Wer. Soc. ii. p. 476.

Skerryvore, the greatest recorded pressure being one of 665 lbs.[1]

From this short and incomplete survey of what has been done by the waves round the Scottish coast during the last two or three hundred years, it is evident that although here and there from local causes, such as the accumulation of sand and shingle, there may have been a slight gain of land, the general result has been a loss. Where the coast is rocky and precipitous this loss may not be measurable, but yet the ruined masses, undermined by the waves, tell their story not less convincingly than where there are historical records of the devastation.

Looking at the general results of marine denudation we are led, on reflection, to perceive that they tend to the formation of a great plain. As it is only the uppermost layer of the sea—the part that is thrown into commotion by disturbance of the atmosphere—which possesses any real power of abrasion, the effect must be not to cut out valleys, but to eat into the land horizontally, and reduce it to a general level under the waves. This tendency is sometimes well illustrated on a small scale on a rocky beach. To the south of Girvan, for example, the Ayrshire coast exhibits between tide marks a smooth level platform of Silurian greywacke, indenting the line of rugged crags that run along high

[1] Stevenson, Trans. Roy. Soc. Edin. xvi. 30.

water-mark. This platform has been cut out of vertical strata, some of which being harder than their neighbours, rise above it into fantastic knobs and bosses. It abounds in cavities lined with sea-weeds and filled with sea-water—each a natural aquarium,

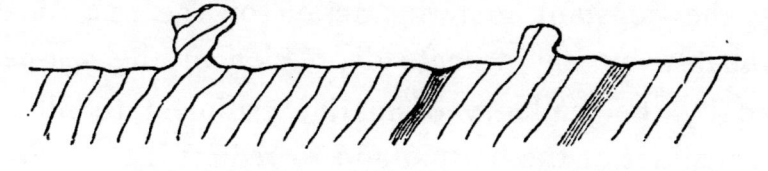

FIG. 3.—INCLINED SILURIAN STRATA NEAR GIRVAN, cut into a plane surface by the sea.

and in some cases, at least, it is evident that these hollows are simply *pot-holes*, like those in the channel of a river, save that the boulders which lie at their bottom have been kept whirling round in the eddies of a vexed tide-way, instead of a rapid brook or river. The level platform, with its hollows and outstanding crags, is a true *plain of marine denudation*, and illustrates in detail a process of which the more gigantic results will be considered in a succeeding chapter.

The sea between the British Isles and the coasts of the Continent from the south of Norway to the north-western headlands of France is so shallow, that an elevation of only 600 feet would convert it all into dry land.[1] And this dry land would present

[1] De la Beche, Theoretical Researches, and Geological Observer.

a wonderfully level contour ; it would have no hills, or rather undulations, so much as six hundred feet high, and its general character would be that of a wide plain sloping gently up to the higher lands on all sides. This area seems, indeed, to be a great plain of marine denudation, cut out of ancient Europe by the constant gnawing action of the sea, aided probably by an ultimate depression of the sea-bed and by the tendency of marine sediment to fill up inequalities of the bottom and to reduce it to a level. But even if subterranean movements came into play we shall perchance not err in attributing mainly to the influence of the waves the actual levelling of the ground and the production of the great submarine plain.

WIND.

As another geological agent capable of making certain minor yet characteristic modifications in the scenery of the country, the wind claims a passing notice here. The influence of high gales in augmenting the destructive force of the sea is of course sufficiently obvious. But the wind also acts directly in the production of terrestrial changes. By prostrating forests it has sometimes given rise to wide barren peat-mosses, by whirling round grains of sand and small pebbles in the crannies of a rock or boulder it excavates little hollows, very much as a river works out pot-holes, and by driving up the sand thrown ashore by the tides, it has converted many a square mile of level ground into tracts of wavy ridge and undulating hollow, like the crests and troughs of a billowy sea. It is when acting in this latter capacity that the wind produces its most familiar effects upon the surface of the ground. Round the Scottish coast-line, wherever prevalent winds blow upon a sandy beach, we usually find a

" Sand-built ridge
Of heaped hills that mound the sea."

Such is the aspect of the wide Tents Muir between the bay of St. Andrews and the mouth of the Tay, the ridges there running in a general sense parallel to each other and to the coast-line. In the Dornoch Firth, along the west side of the islands of Coll and Tiree, in Macrihanish Bay, Cantyre, and along the Ayrshire shores for fully fifteen miles between Irvine and Ayr, as well as on many other parts of the coast, the characteristic results of wind-action on loose sea-sand may be seen. The inland advance of the sand has sometimes destroyed large tracts of fertile land. Thus along the shores of the Moray Firth the old barony of Culbin has been wholly obliterated. " I have wandered for hours," says Hugh Miller, " amid the sand-wastes of this ruined barony, and seen only a few stunted bushes of broom, and a few scattered tufts of withered bent, occupying, amid utter barrenness, the place of what, in the middle of the seventeenth century, had been the richest fields of the rich province of Moray."[1] So on the Aberdeenshire coast many a fair field has disappeared. "The parish of Forvie," says Pennant, "is now entirely overwhelmed with sand, except two farms. It was in 1600 all arable land, now covered with shifting sands, like the deserts of Arabia, and no vestiges remain of any buildings except a small fragment of the church."[2]

[1] Sketch Book of Popular Geology, p. 13.
[2] First Tour, p. 144.

CHAPTER IV.

ACTION OF GLACIERS AND ICEBERGS.

THE agencies of waste described in the preceding pages are still at work around us, and their hand can be seen in every landscape in the country. But in tracing back the origin of our present scenery we are soon led to remark that a vast amount of superficial change must be due not to these agents, but to some other which is no longer busy within our borders. In the northern half of the British Isles, as well as generally throughout northern Europe and America, there is abundant evidence that the surface of the land has been thus specially modified. Had the denudation of these regions been the work only of the sea and ordinary sub-aërial waste, their general outlines would have been more rugged than they are. The valleys would oftener have had steep craggy sides, the hills would have been sharper in form, and more deeply cleft with gullies and ravines. There is on the contrary a general smoothness of contour, characteristic alike of hill and valley, pointing to some abrading agency which has in large measure worn off the old roughnesses, and given a flowing outline to the ground. This agency is now becoming recognised as that of land ice, which must once have thickly covered the whole country, moving steadily downwards to the sea, and in its journey grinding away, smoothing, and striating the surface of the rocks. After long years of doubt and discussion, geologists are at length led to believe that during a comparatively recent geological period the whole of the northern half of Great Britain was cased in ice as North Greenland is to-day. The evidence for this inference will be given in a subsequent chapter, and that it may be the better followed we may glance for a little at the nature of the geological results which are produced by ice at the present time.[1]

The aspect of North Greenland probably affords us a close parallel to the state of Scotland during the climax of those geological changes which were comprised within what is called the Glacial or Drift Period. The interior of that tract of country is covered with one wide sheet of snow and ice, which, constantly augmented by fresh snow-falls, moves steadily downward from the axis of the continent

---

[1] See a Memoir by the Author on "The Phenomena of the Glacial Drift of Scotland," published as Part II. of the Transactions of the Geological Society of Glasgow.

to the eastern and western shores. This vast *mer de glace* protrudes in some places for several miles into the sea, where it breaks up into huge masses which drift away as icebergs. Inland it sweeps league after league in one interminable glacier, broken only here and there by some black hill top or mountain peak that rises as an island out of the snow. Its thickness, too, must be very great. Between latitudes 79° and 80°, Dr. Kane found it abutting as a solid glassy wall, which rose 300 feet above the sea level, and sank to an unknown depth below it. And this massive ice-sheet is ever slowly and persistently creeping down to the sea. It covers the face of the country, filling up the valleys, mounting over the hills, and pressing with constant resistless force upon all the rocks over which it marches.

The geological effects of a mass of ice moving steadily from a higher to a lower level are most important. Blocks of stone, either loosened from the parent mass by frost, or broken off by the moving glacier, are jammed in the ice and pressed along the rocky bed or sides of the valley. The stones, gravel, and sand thus act the part of files; they scratch and score the hardest rocks, and are themselves striated by the same process. Hence year by year as this rasping and polishing goes on there cannot fail to be a considerable loss of rock. In the Swiss valleys this loss is impressively shown at the lower end of a glacier, where the river, which leaps into the light, is charged with fine mud, the result of the grating of the sand and stones under the ice upon the rocks of the valley.

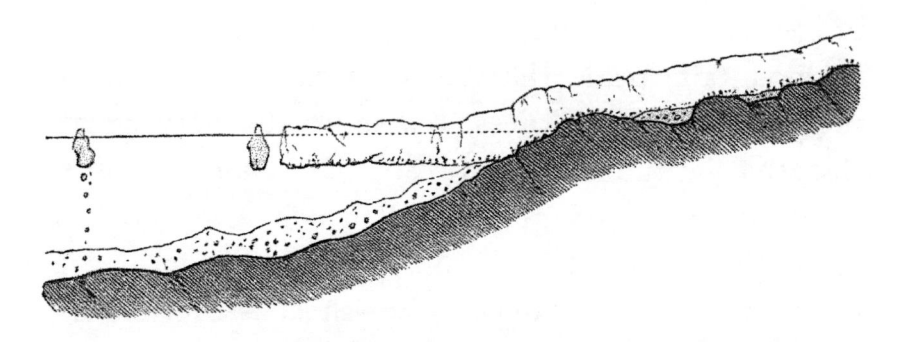

FIG. 4.—SECTION OF A SHEET OF LAND-ICE, going out to sea, breaking off there into bergs, and forming *boulder-clay*, partly on land, and partly in the sea.

If a single glacier, descending far below the snow-line, as in Switzerland, can grind down, polish, and score the rocks of its channel, it is easy to see how vast and constant must be the erosion carried on by so huge a mass of ice as that which creeps over the whole of North Greenland from mountain top to seashore. Could we strip off this icy mantle, we should find the surface of that country worn into rounded and flowing outlines, its valleys and hills smoothed, and its rocks covered with ruts and grooves running in long persistent lines, and marking the direction of the march of the ice. It would probably bear a close resemblance to the surface of Norway, and even as we shall see to many parts of the hilly tracts of the British Isles.

One characteristic result of the ice-action of the Glacial Period was the scooping out of hollows in solid rock now holding water and forming lakes. To my old and valued friend, Professor A. C. Ramsay, belongs the merit of having first suggested and worked out this idea. He has shown that the innumerable rock-enclosed basins of the northern hemisphere do not lie in gaping fissures produced by underground disturbances, nor in areas of special subsidence, nor in synclinal folds of the strata, but that they are true hollows of erosion. Neither rivers nor the sea could scoop out such hollows. The only known agent by which the work could have been done is *land-ice*.[1] We shall note in a succeeding chapter how well this explanation is borne out by the innumerable lakes, lochans, and tarns that are scattered so widely over the country.

In a glacier valley with rocky declivities on either side, the frosts, thaws, and rains of every year loosen

[1] Reference should be made to Professor Ramsay's memoir on this subject. *Quarterly Journal Geo. Soc.* vol. xviii. p. 185. See also his summary in his "Physical Geology and Geography of Great Britain." Since his memoir was published it has been violently assailed both in this country and on the continent. I have seen no argument, however, to shake my belief in my friend's views and the additional experience of the interval that has passed since their publication has enabled me still further to test their truth. See also the Memoir on the Glacial Drift of Scotland, already quoted, p. 86.

large quantities of débris. As the glacier threads its way down the valley, its surface gets discoloured with the mud washed from the slopes, piles of rubbish collect along its sides in long lines called *moraines*, which are slowly borne onward with the ice, till at last when the glacier melts, they are thrown down in confused heaps that are apt to be washed away by the river as fast as they are deposited. Where, however, these rubbish heaps remain they form a more or less continuous rampart along or across the valley. And by the intermitted recession of the glacier successive lines of such barrier-mounds may be thrown down, each line marking a pause in the retreat of the ice. In many Scottish valleys, as will be afterwards pointed out, these moraine mounds remain singularly fresh, still damming up the waters as they did when the ice drew back from them, and still strewed with huge boulders that were once carried on the ice from the higher recesses of the hills and scattered on the heaps of rubbish where they still lie.

Land-ice is thus a most powerful geological agent in new-modelling the surface of the earth. But in northern latitudes the same ice which plays so important a part on the land descends to the sea-level, and breaking off there into icebergs, performs a new series of feats on the great deep. These ice-islands carry with them any soil or rock rubbish that may have fallen upon them from inland cliffs, while they formed part of the ice-sheet of the country. The débris so borne off is, of course, thrown down upon the sea-bottom as each berg melts away after a voyage of perhaps several thousand miles. Year by year whole fleets of these bergs are sent southwards in the arctic regions, so that the bed of these northern seas must be plentifully strewed with earth and boulders. As only about a ninth part of a mass of ice appears above the water on which it floats, the bulk of many bergs must be enormous. One rising two hundred feet above the waves—not an uncommon height—must have its bottom sixteen hundred feet below them. Deeply seated in the water, they are acted on much more by marine currents than by winds. Hence, they are sometimes seen careering through a frozen sea in the teeth of a tempest, breaking up the thick ribbed ice before them with a noise like the loudest thunder, yet with as much apparent ease as a ploughshare cuts the loam. Every winter, crowds of these bergs are firmly fixed in the frozen sea of the arctic regions, and when summer comes the united mass drifts southwards towards Newfoundland. Vast sheets of ice, larger sometimes than the whole of Scotland, with ice-hills rising two or three hundred feet above the sea-level and sinking two thousand feet or more below it, are thus borne by the ocean currents into warmer latitudes, where they break up and disappear.

When such current-driven masses grate along the sea-bottom they must tear up the ooze and break down and scratch the rocks. In the course of long ages a submerged hill or ridge may get its crest and sides much bruised, shorn, and striated, and

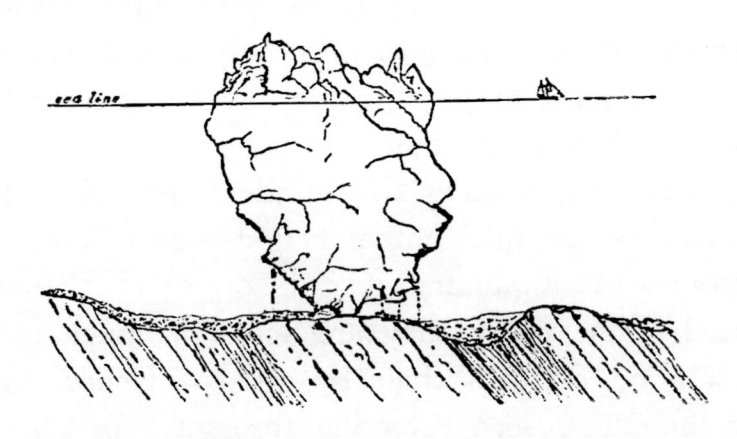

FIG. 5.—ICEBERG grating along sea-bottom, and depositing mud and boulders.

the sea-bed generally may be similarly grooved and polished, the direction of the striation being more or less north and south according to the prevalent tread of the drifting ice.

Besides the land-ice which breaks off into bergs, the *ice-foot* or frozen margin of the sea deserves notice as a geological agent. It receives huge piles of rubbish disengaged by frost from the cliffs that overhang it, and when broken up by storms, masses of it, loaded to the full with earth and stones, are borne seawards, where as they melt, they drop their burden to the bottom ; so that both from bergs and coast-ice the floor of the northern seas must be receiving constant additions of mud, sand, and erratic blocks. These modern changes help to throw not a little light on certain wide-spread deposits of clay and boulders which largely influence the scenery of the low grounds of Scotland.

It appears then that at the present day there are a number of agencies unceasingly at work in altering the present outlines of the dry land. Rains, springs, and frosts, are slowly widening and deepening our valleys, and the sea is eating into the coast and labouring to plane down the country to a general level under the waves. We learn, moreover, that in a comparatively recent period the surface of the greater part of the British Isles has been greatly modified by the erosive action of land-ice and icebergs, and evidence is not wanting that ice has played a like part during earlier geological periods. In the revolutions which befall the surface of the earth it is the province of the sea to wear away the shores and to continue the work of destruction until the land is levelled down. The result of this waste is the formation of a *plain of marine denudation*. The task of rains, springs, streams, frosts, and glaciers, is to deepen the hollows of such a sea-worn surface after it has been upheaved into dry land, carving out of it systems of wide and deep valleys and leaving the intervening parts to rise up as chains of hills and mountains. To these various forms of denudation, working during a long succession of ages, the present scenery of the country must, I believe, be mainly attributed.

The only other great agency which can have had a share in this work is subterranean movement. But it has already been pointed out that the valley systems cannot have been formed merely by that cause, but must have been in the main carved out of the land by the action of the atmospheric powers of waste. So far as we can venture to affirm positively in the matter, it would seem that the chief result of underground movements has been to raise or depress different districts, in the one place elevating a sea-bottom into dry land and thereby exposing it to the erosive action of the elements, in the other case carrying down a land surface beneath the sea and so protecting it from further waste. Such movements may no doubt have been accompanied by local and paroxysmal disturbances rending the ground into those cracks and chasms familiar in the history of earthquakes. But even such more violent exhibitions of the power of the deep-seated forces may have produced no more extensive or lasting effects on the surface than the most powerful earthquakes in historic times have done. More probably the movements were general, gradual, and long continued, the lands slowly rising foot by foot above the sea level, or sinking with like gentleness below it. It is not to be forgotten, however, that these movements, though tranquil, may have been very unequal, even over comparatively limited districts, one part of the country rising or sinking to a much greater extent than other portions. Such inequalities could not fail to have produced great changes upon the drainage of the country, and consequently upon the progress of the excavation of the valleys.

The history of the superficial changes of a country might be theoretically summed up thus. The bed of the sea, which for the sake of illustration may be supposed to be a wide plain of marine denudation—the worn remnant of an old land—is slowly raised above the waves. There is a point or line where the elevation is greatest, and from which the ground slopes down to the sea-level. Perhaps the elevatory force shows itself in the upheaval of one or more anticlinal folds. But whether in one great dome-shaped mass, or in a long ridge, or in several ridges with parallel dividing hollows, the slopes are probably gentle, and the elevation on the whole a quiet protracted process. No sooner does the rock appear above the sea than it is attacked by the waves, and unless the rate of elevation is more rapid than that of the marine waste, the rising area can never get above the sea-level. But not only is it a prey to the breakers, its surface begins to be carved out by the atmospheric agents. That surface is not a mere dead level, so that when rain falls upon it drainage necessarily sets in from the highest parts down to the shores. The rain gathers into runnels, following the inequalities of the sea-worn slopes, and widening into brooks and rivers; or the moisture falls in the form of snow, and glaciers-grind a path for themselves from the high grounds to the shore. Thus begins the scooping out of a system of valleys diverging from the higher parts of the rising land. These depressions are slowly dug deeper and wider, until at last the ancient elevated sea-bed is worn into a system of hills and mountains, valleys and glens. The land thus modelled may remain stationary for a vast interval, but in the end it descends again beneath the sea, is covered over with newer deposits, and its highest mountains perhaps buried beneath piles of their own ruin, worn from them by the sea, as they slowly sink under its waves. A subsequent elevation of this area into dry land exposes these later accumulations to a similar waste, and a new series of denudations is begun by the rains, streams, ice, and the sea, new valleys are excavated, and new hills are left standing out from them. By such a process of ceaseless change thus summarily stated, carried on during many successive geological periods, the present scenery of our country appears to have been produced.

In the succeeding chapters I propose to attempt to trace the origin of the present scenery of Scotland to these causes. Though we may understand the general character of a process in nature, it is sometimes by no means an easy task to follow out the details of its successive stages, and this difficulty is vastly increased when the process was completed long ago, and when, in consequence, its memorials are obscure and incomplete. Such, however, is the task now before us. In entering upon it, I feel that it is possible in the meanwhile merely to grope the way. The only light which can be taken with us is that of existing nature. Without its help all would be utter darkness, but under its direction we may be enabled to advance some little way. I shall endeavour to lay before the reader the facts on which each inference is based, that he may judge how far there is actual evidence from which to restore in imagination certain ancient conditions of our country.

There are in a broad sense three distinct types of scenery in Scotland, each due to certain distinguishing geological features. Although they are intimately linked with each other and show the working of the same great geological processes, it will be, perhaps, the most intelligible mode of treatment to describe each separately. There is, therefore: 1st. The scenery of the Highlands; 2nd. The scenery of the broad tract of high ground between St. Abb's Head and Portpatrick, sometimes called the Southern Highlands, and to which for the sake of clearness I shall refer as the Southern Uplands: and 3rd. The scenery of the Midland Valley, of the Tay, Forth, and Clyde. My account of the Highlands will necessarily be the longest, since it will not be needful to repeat in detail under the other two divisions the explanation of the formation of valleys and lakes, and the history of the glacial period.

CHAPTER V.

THE SCENERY OF THE HIGHLANDS.

GEOLOGICAL STRUCTURE OF THE DISTRICT.

A LINE drawn from the Coast of Kincardine in a south-western direction across the island to the Mull of Cantyre divides the Highlands from the central valley of the Tay, Forth, and Clyde. It coincides accurately with the boundary between the old crumpled gneisses and schists of the northern half of the kingdom and the conglomerates, red sandstones and trap-rocks of the broad Lowland Valley. Thus it is both a good geological line and a well-marked limit for two very different types of scenery. To the southeast lie the gently undulating hills and wide agricultural plains of the Old Red Sandstone; to the north-west a sea of mountains rolls away to Cape Wrath in wave after wave of gneiss, schist, quartz-rock, granite, and other crystalline masses.

This mountainous tract, forming the Highlands of Scotland, is bordered on the north-east with a belt of Old Red Sandstone, which gives rise to a strip of fertile land along the shores of the Moray Firth. On the west it is flanked with a chain of islands formed chiefly of igneous rocks of secondary or tertiary age. But within this outer frame the enclosed mountainous region is made up of older Palæozoic strata which, over the greater part of the surface, have passed from their original character of ordinary marine sediment, —sand, gravel, silt, and mud—into gnarled crumpled crystalline masses. A few years ago these rocks presented a wild chaos of disorder, out of which it appeared hopeless to think of ever making any recognisable geological structure. An upward order of succession had, indeed, been determined by the late Dr. Macculloch and Mr. Hay Cunningham among the rocks of the west of Sutherland, but no one thought of applying it to the elucidation of the structure of the rest of the Highlands. It was not until 1855 that Sir Roderick Murchison, aided by Mr. Peach's discovery of Lower Silurian fossils in the limestones of Duirness, established the true relations of the rocks of Sutherland and Ross, and obtained the key with which he has revealed the structure of the Scottish Highlands—a discovery the importance of which it is hardly possible to over-estimate.[1] At

[1] See his papers, and new Geological Map of the Highlands, published in the Quarterly Journal of the Geological Society for 1858. The section across Sutherland, on the map accompanying the present volume, shows the order of superposition.

the bottom of all lies a gnarled highly crystalline gneiss, called by Sir Roderick the Fundamental Gneiss,[1] inasmuch as he has shown it to be the oldest rock in the British Isles, and the foundation on which all the others are built up. Resting on the upturned worn edges of this truly primæval rock rise huge masses of gently-inclined red sandstone and conglomerate, supposed to be of the same geological age as the Cambrian rocks of Wales. These strata, formerly described and mapped as Old Red Sandstone, form the lofty pyramidal mountains on the Atlantic borders of Sutherland and Ross, and are confined to the north-west of the island. They are overlaid with thick quartz rocks and bands of limestone which pass under flaggy micaceous gneisses and schists, and contain in their under-portions Lower Silurian fossils. Hence we learn that the rocks which overlie the red Cambrian sandstones, even though they are crystalline and wholly changed from the soft loose strata in which they were laid down on the sea-bed, yet do not belong to the age of chaos, as used to be supposed, nor to an epoch anterior to the creation of life, but are really the equivalents in time of the Lower Silurian rocks of the south of Scotland, of Cumberland, and of Wales. A reference to the long section on the map from Skye to the Cheviot Hills will show

[1] He has also called it "Laurentian" from its probable identity with a gneiss occupying a similar position in Canada.

the reader that it is these metamorphosed Lower Silurian strata which form the greater part of the Highlands, and that they are arranged in great folds, whereby the lower portions of the series are brought up to the surface again and again.[1]

The strata of sand and mud which accumulated to a depth of thousands of feet over the sinking floor of the old Silurian ocean have been crumbled up into endless folds and puckerings of which, as may be seen on the map, the long axis, or *strike*, runs generally in a north-easterly and south-westerly direction. When the wind blows freshly from the north-west the sea is roughened with long broken lines of wave, stretching from south-west to north-east, and rolling in towards the south-east. So over the Scottish Highlands the gneissose and schistose rocks have been tossed, as it were, by a long swell from the north-west into numerous wave-like plications that follow each other, fold after fold, and curve after curve, from Cape Wrath to the Lowland border.

Thus, instead of a scene of undistinguishable chaos, the Scottish Highlands are found to be governed by the same laws of structure as other hilly regions formed of stratified rocks. We are not left, therefore, to speculate wholly in the dark upon the origin

---

[1] See the details of this structure, given in a Memoir on the subject by Sir Roderick Murchison and the author in the *Quarterly Journal of the Geological Society*, vol. xvii. 171.

of the Highland mountains and glens, as if they had been upheaved among rocks of which the mode of formation and the actual structure were alike unknown. If we were still vaguely surmising that the gneiss and the schist had been thrown down upon the floor of a primæval thermal ocean, and had been broken up when that hardened floor was upheaved into the first dry land, there might be some excuse for a belief that the ancient convulsions had been the means of throwing up our northern mountains and tearing open their glens. But when these rocks are discovered to be only a modification of ordinary sedimentary deposits, and to reveal their geological age by their enclosed organic remains, all such vague conjectures must cease, and the rocks must be examined and determined by the ascertained laws which govern the arrangement of masses of stratified rocks.

Connexion of Geological Structure and Scenery.—When it is said that the rocks of which the Highlands are made have been thrown into oft-repeated foldings it might be supposed that these undulations of the strata have given rise to the present contour of the surface, each mountain-chain corresponding with an arch or *anticlinal axis* of the gneiss and schists, and each line of strata or glen with a trough or *synclinal axis*. On the contrary, the very reverse of this arrangement is often found to hold. What in a geological sense are basins or troughs, frequently rise into rugged and lofty mountains, while the arches, on the other hand, are occupied by deep valleys. A striking example of this feature is to be found in Ben Lawers. That wide-based broad-shouldered moun-

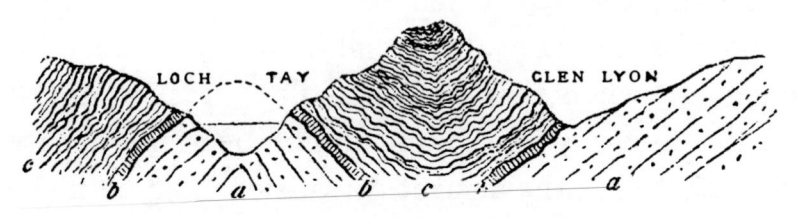

FIG. 6.—SECTION OF BEN LAWERS.
a Quartzose rocks. *b* Limestone. *c* Schistose rocks.
(See also the section No. 2 in the Map.)

tain rises from the valley of Loch Tay on the one side, and sinks into Glen Lyon on the other. It forms thus a huge dome-shaped mass between two deep valleys. But instead of owing this form to an upward curving of the schists, it actually lies in a basin of these rocks which dip underneath the mountain on the banks of Loch Tay, and rise up again from its further skirts in Glen Lyon. Thus Ben Lawers is in reality formed of a trough of schists, while the valley of Loch Tay runs along the top of an anticlinal arch. Hence that which, in geological structure, is a depression has by denudation become a great mountain, while what is an elevation has been turned into a deep valley.

The present heights and hollows of the Highlands, therefore, are not to be traced to any of the original convolutions of the old crystalline rocks. Nor can they be assigned, as they often are in a popular way, to grand primeval eruptions of granite. That rock, it is true, covers a considerable space in the Highlands, and rises up among the highest mountain groups. But there are also wide spaces of low ground abounding in granite. The long lonely Moor of Rannoch, for instance, lies in large measure on granite; while the range of mountains that bounds its south-eastern margin consists, not of granite, but of quartz rock. Indeed, there is good reason to believe that granite is not an igneous rock in the ordinary sense, but that instead of bursting through and upheaving the gneiss and schist, it is itself only a further stage of the metamorphism of these rocks.

Until the Highland tracts are surveyed in minute detail, it will not be possible to ascertain how far they are traversed by lines of fault, nor to what extent such features have shewn themselves at the surface, and have served to guide the excavation of the valleys. After a long and detailed examination of the contorted rocks of the Silurian uplands of the southern counties, I have been led to believe that the faults and the folds of the strata on the whole have had only a secondary influence in originating the present irregularities of the surface. And this is probably the case also with the contorted and metamorphosed Silurian rocks of the Highlands. The

Great Glen, to be afterwards more specially referred to, is one of the few examples yet known to me in Scotland of the coincidence of a line of fault with that of a long valley.

While, however, the character of Highland scenery has not been mainly determined by subterranean movements, it is nevertheless true, as will be afterwards pointed out, that the forms into which the rocks were thrown by the contortions and dislocations have, in many cases, materially guided the powers of waste in the long process of denudation. The larger features,—hill ranges and lines of valley,—have sometimes had their general trend determined by the direction of the anticlinal and synclinal foldings of the strata. The minor details, which give individuality to the forms of cliff and crag and mountain, have likewise been dependent upon the lithological character of the rocks. But alike in the greater and the lesser elements of the scenery, there has been a presiding power of erosion, which, though its working might be modified by local circumstances, has laid its finger on every rood of the surface, and has carved out for itself the present system of glen and mountain, valley and hill.

It cannot be too strongly impressed upon the mind that the existing surface of the area of the Highlands is far from being the same as that which existed when the rocks were squeezed, crumpled, and broken.

These changes went on beneath the surface under a vast thickness of rock, which has since been worn away. There is now no trace of the original effects produced by these underground movements upon the exterior of the earth's crust. If they ever made any show there at all (which seems to me by no means certain) it was effaced long ages ago. The present mountain-tops have been revealed to the light of day only after the removal of hundreds and thousands of feet of solid rock, under which they once were buried. Thus one of the first lessons we learn in attempting to restore former aspects of Highland scenery is an impressive one of vast waste. There is not a single mountain or glen by which this lesson is not brought home to us. Each is in itself a monument of denudation telling everywhere of the work of rains and streams, sea, frost, and ice, but giving no clue to the effects of primeval earthquakes.

ANCIENT TABLE-LAND OF THE HIGHLANDS.

IF one would grasp at once the leading features of Highland scenery, let him betake himself to some mountain-top that stands a little apart from its neighbours, and looks over them into the wilds beyond. A better height could not be chosen than the summit of Ben Nevis. None other rises more majestically above the surrounding hills, or looks over a wider sweep of mountain and moor, glen and corry, lake and firth, far away to the islands that lie amid the western sea. In no other place is the general and varied character of the Highlands better illustrated. And from none can the geologist, whose eye is open to the changes wrought by subaërial waste on the surface of the country, gain a more vivid insight into their reality and magnitude. To this, as a typical and easily accessible locality, I shall have occasion to refer more than once. Let the reader, in the meantime, imagine himself sitting by the side of the grey cairn on the highest peak of the British Isles, watching the shadows of an autumnal sky stealing over the vast sea of mountains that lies spread out, as in a map, around him. And while no sound falls upon his ear, save now and then a fitful moaning of the wind among the snow rifts of the dark precipice below, let him try to analyse some of the chief elements of the landscape. It is easy to recognise the more marked heights and hollows. To the south, away down Loch Linnhe, he can see the hills of Mull and the Paps of Jura closing in the horizon. Westward, Loch Eil seems to lie at his feet, winding up into the lonely mountains, yet filled twice a day with the tides of the salt sea. Far over the hills, beyond the head of the Loch, he looks across Arisaig, and can see the cliffs of the Isle of Eigg and the dark peaks of Rum, with the Atlantic

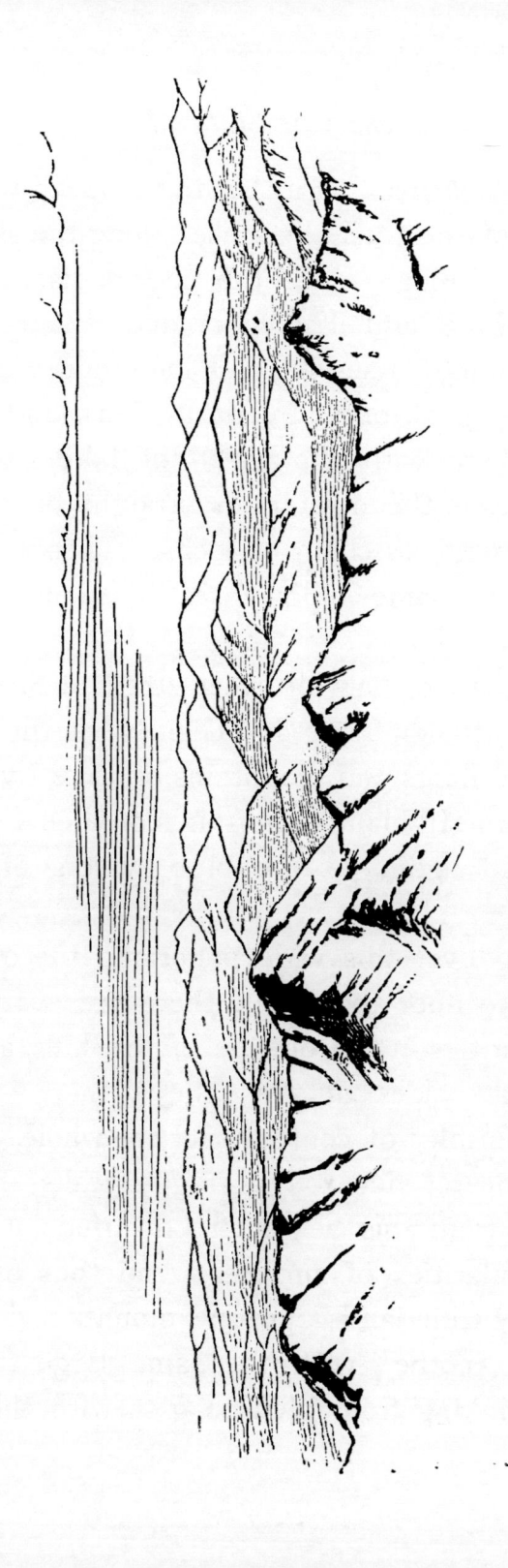

FIG. 7.—VIEW FROM TOP OF BEN NEVIS.

gleaming below them. Further to the north-west the blue range of the Cuchullins rises along the sky-line, and, then, sweeping over all the intermediate ground, through Arisaig and Knoydart and Claurannald's country, mountain rises beyond mountain, ridge beyond ridge, cut through by dark glens, and varied here and there with the sheen of lake and tarn. Northward runs the mysterious straight line of the Great Glen, with its chain of lochs. Thence to east and south the same billowy sea of mountain-tops stretches out as far as eye can follow it—the hills and glens of Lochaber, the wide green strath of Spean, the grey corries of Glen Treig and Glen Nevis, the distant sweep of the moors and mountains of Brae Lyon and the Perthshire Highlands, the spires of Glen Coe, and thence round again to the blue waters of Loch Linnhe.

In musing over this wide panorama the observer cannot fail to note that while there are everywhere local peculiarities in the outline of the hills, and the shapes of the sides of the valleys, there is yet a general uniformity of contour over the whole. What seem at a nearer view rough craggy peaks and pinnacles, seen from this height are dwarfed into mere minor irregularities of surface. And thus over the whole of the wide landscape one mountain ridge appears after another, with the same large features, rising and sinking from glen to glen with the same smoothed summits, broken now and again where from some hidden valley a circular corry or craggy cliff lifts itself bare to the sun.

Much has been said and written about the wild tumbled sea of the Highland hills. But, as he sits on his high perch, does it not strike the observer that there is after all a wonderful orderliness, and even monotony, in the waves of that wide sea? And when he has followed their undulations from north to south, all round the horizon, does it not seem to him that these mountain-tops and ridges tend somehow to rise up to a general level, that in short there is not only on the great scale a marked similarity of contour about them, but a still more definite uniformity of average height? (See Figs. 7 and 8.) To many, who have contented themselves with the bottom of the glen, and have looked with awe at the array of peaks and crags overhead, this statement will doubtless appear incredible. But let any one get fairly up to the summits, and look along them, and he will not fail to see that the statement is nevertheless true. From the cairns on Ben Nevis this feature is impressively seen. Along the sky-line the wide sweep of summits undulates up to a common level, varied here by a higher cone, and there by the line of some strath or glen, but yet wonderfully persistant round the whole panorama. If, as sometimes happens in these airy regions, a bank of cloud with a level under-surface should descend upon the mountains, it will be seen to touch summit after summit, the long line of the cloud defining like a great parallel-ruler the long level line of the mountain-ridges below. I have seen this feature brought out with picturesque vividness over the mountains of Knoydart and Glen Garry. Wreaths of filmy mist had been hovering in the upper air during the forenoon. Towards evening, under the influence of a cool breeze from the north, they gathered together into one long band that stretched for several miles straight as the sky-line of the distant sea, touching merely the higher summits and giving a horizon by which the general uniformity of level among the hills could be signally tested. Once or twice in a season one may be fortunate enough to get on the mountains above such a stratum of mist, which then seems to fill up the inequalities of the general platform of hill-tops, and to stretch out as a white phantom-sea, from which the highest eminences rise up as little islets into the clear air of the morning.

The long level line of the Highland mountain-tops may be seen not only from some commanding eminence among the mountains themselves, but perhaps even more markedly from the lower country outside. From the isles of Skye and Eigg, for instance, the panorama between the heights of Applecross and the Point of Ardnamurchan shows in a most impressive manner the traces of the old table-land. So too when

FIG. 8.—VIEW OF THE HIGHLAND MOUNTAINS FROM ABOVE GREENOCK.

the observer ascends the ridges behind the town of Greenock, the hill-tops of the opposite Highlands, between Inellan and the head of Loch Long, stretch out before him in a long and almost straight line.

What does this general uniformity of level mean? It plainly has nothing to do with geological structure. On the contrary, a cursory examination will show that the abraded edges of the highly inclined strata of gneiss and schist, rise to the surface along and across the hill-tops, and that these strata must once have been prolonged upward to complete the folds which only remain in part. There must consequently have been a vast deal of denudation along these heights, and an enormous mass of rock must have been worn away from the tops of the Highland mountains. The uniformity of level among these summits arises not from any structure in the rocks, but from the activity of the powers of waste. If the foregoing observations on the functions of the different denuding agencies were just, it follows that the long flat surfaces and the common average height of so many of the Highland mountains are due to the levelling power of the sea. In other words, these mountain tops are parts of a great undulating plain or table-land of marine denudation.[1]

[1] Let me guard the reader against misconception arising from the use of this term *plain*. I do not at all mean to assert that the area of the Highlands was ever so levelled by the sea as to approach to anything like the flatness of a meadow. The marine denudation probably

That this plain can now be traced only in fragments, that it is cut down by wide straths and deep glens, and that its surface has been greatly and unequally lowered in many places, need not be matter of surprise. When we come to reflect on its extreme antiquity and on the long cycle of geological revolutions that have rolled over the country since then, the wonder rather is that any trace of the plain should remain at all. The valleys which now intersect it, as we shall immediately see, have probably been dug out of it by the agencies of denudation. If, therefore, it were possible to replace the rock which has been removed in the excavation of these hollows the Highlands would be turned into a wide undu-

went on during many oscillations of level, and the general result would hence be the production of a great table-land, some parts rising gently to a height of many hundred feet above other portions, yet the whole wearing that general tameness and uniformity of surface characteristic of a table-land where there are neither any conspicuous hills towering sharply above the average level, nor any valleys sinking abruptly below it. It would require much more detailed surveys than I have yet been able to make, to warrant me in fixing the probable average height of this ancient Highland table-land above the sea. Its highest parts were probably at least as lofty as the present summits of Ben Nevis and Ben Muick Dhui ; its lowest hollows were in all likelihood not less than 1,000 feet above the present sea-level. Nor must we forget that this table-land, since the time of its formation, has doubtless shared in the numerous upheavals and depressions of the country. These movements, though they may have destroyed in part the continuity of the plain, have left it still wonderfully distinct. Hence it is not so difficult to understand why there should have been a wide table-land, as to conceive how it has not been in great measure or wholly effaced. Compare Prof. Ramsay's description of Wales. "Lectures," p. 140.

lating table-land, sloping up here and there into long central heights and stretching out between them league after league with a tolerable uniformity of level. And in this rolling plain we should find a restoration of the bottom of a very ancient sea.

Is there any evidence of the geological age in which that sea covered the Highlands? for if this can be obtained some light may perhaps be cast on the subsequent changes that have turned the ocean-floor into a wide region of hill and valley. The evidence which can be gleaned is far from conclusive, yet it is not without value and interest.

The metamorphism of the strata that form the Highlands must have taken place between the deposition of what are known as the Llandeilo or Caradoc rocks and that of the Lower Old Red Sandstone. It is thus in all likelihood of Upper Silurian date. Wherever the Old Red Sandstone lies on these Highland rocks there is satisfactory proof of great denudation, for the conglomerates rest upon the worn edges of the inclined or vertical gneiss and schist, of whose ruins, indeed, they are composed. So that the rocks of the Highlands were not only metamorphosed, but largely worn away before the Old Red Sandstone was laid down upon them. The northern and eastern coasts of the county of Sutherland are fringed with a broken band of the Lower Old Red conglomerates and sandstones, which form a chain of rounded, craggy, conical hills between Golspie and Helmsdale, rising sometimes to nearly 2,000 feet above the sea, and presenting the abraded ends of the strata towards the interior. It is impossible to look at these brown hills without being convinced that they remain as a mere fragment of a great sheet of conglomerate and sandstone which stretched away into the interior. But as if to make this point quite certain, in the very heart of the county among the gneisses and granites rise up the two mountains of Ben Griam, the higher being 1,935 feet over the sea-level. These are cakes of conglomerate lying on the worn surface of the crystalline rocks out of the detritus of which they have been formed. They are thus only remnants of a once wide deposit. So again, along the northern shores, patches of the same kind are found from the borders of Caithness to Roan Island, sometimes in little outliers standing high among the inland hills. Hence it must be inferred that a large part, if not the whole, of the county of Sutherland was once covered with a sheet of Old Red conglomerate, of which there are now left only a few relics capping some of the heights of the interior and fringing the coast-line. The same deposit runs southward from Sutherland along the eastern coasts of Ross and the shores of the Moray Firth. It stretches up the valley of the Great Glen[1]

[1] This valley, therefore, would seem to be at least as old as the Lower Old Red Sandstone.

and rises in Mealfourvonie to a height of 2,700 feet. Thence it sweeps eastwards along the seaboard of the counties of Inverness, Nairn, Elgin, Banff, and Aberdeen, and detached portions are found thirty or forty miles in the interior. In this district also it can hardly be doubted that these denuded outliers indicate that the Old Red Sandstone once spread over a wide area of the northern Highlands.

Along the southern border of the Highlands the evidence is less obtrusive, but perhaps not less definite. From sea to sea the Highland mountains are there flanked with the Old Red Sandstone, sometimes in low rolling plains that creep up to the base of the hills, but sometimes, as in the Braes of Doune, rising into long heathery heights, that form a kind of outer rampart to the main mass of the Highlands. Even from a distance the stratification of the conglomerates and sandstones of these uplands can be easily traced, the beds presenting their denuded, truncated ends towards the mountains to which they evidently at one time were prolonged, and from the waste of which they were formed. At Callander, and elsewhere along the border-line, it appears that the red sandstone series has been faulted down against the older rocks to the north-west. Hence to understand the original relation of the formations in this district we must in imagination undo the effect of the fault, and upraise again the depressed sandstones and conglomerates for perhaps many hundred feet. When this is done we see, from their abrupt, truncated ends looking to the north-west, that these strata cannot but have stretched for some way over the mountains.

And thus, by piecing together the evidence furnished by the Old Red Sandstone along the borders and in the interior of these high grounds, we obtain a strong probability that the great denudation which levelled the old Highland table-land, began in the Upper Silurian or at the commencement of the Old Red Sandstone period, and that, as a result of that denudation, a thick pile of conglomerate and sandstone was made out of the waste of the older rocks, and laid down over many a square league of the Highlands. The higher mountain ranges, such as the central chain of the Grampians, perhaps rose as islands out of these ancient waters, after the lesser heights had been ground down and buried beneath their own ruins. From the thickness of the conglomerates it may be inferred that the denudation went on during a slow submergence of the land—a condition eminently favourable for marine abrasion. For this downward movement, ever bringing new zones of land within the power of the breakers, could not but greatly help the sea in its task of planing down the irregularities of the sinking country. It may have ceased before the higher summits were reached by the waves. The area worn away by the marine denudation would thus not be a mere flat plain, but would slope gently under the sea away from these unsubmerged high grounds. Yet from the slowness of the subsidence the angle of this inclined sea-bed might be so gentle as to appear to the eye almost a level surface. And this feature is found still to mark the Highland mountain-tops. A line connecting summit with summit across the Grampian chain, from north-west to south-east, would show a very gradual seaward slope on either side until it reached the outer edge of the high grounds, where it would descend rapidly into the plains.[1]

[1] From the top of Ben Muick Dhui (4,300 feet) north-westward to the crest of the hills overlooking the Moray Firth (say 1,300 feet above the sea-level), is a distance of about 28 miles, and the angle of descent would be not more than about 1 in 50. From the same central elevation to the south-eastern verge of the Highland mountains, the angle would be not quite so much, as the distance is rather more, and the average height of the broad table-land there is greater than on the opposite border. These slopes would not be so steep as some railway gradients now in use. Yet they are perhaps exaggerations of the original average declivity from the central ridge of the Highlands, for the higher parts, such as the Ben Muick Dhui and Ben Nevis groups of mountains, may have risen high above the waves that were planing down the surrounding table-land. Thus the gentle seaward slope may have begun at the foot of groups of hills which had suffered from an earlier marine denudation, and were now undergoing a slow erosion by the atmospheric agencies. For the sake of simplicity I have referred above only to that marine denudation which planed down the Highland mountain-tops previous to and during the deposition of the Lower Old Red Sandstone. But the levelling of the table-land may have gone on during many subsequent geological periods, for the overlying conglomerates and sandstones have been themselves in large part worn away. (See the section in Prof. Ramsay's Lectures, p. 140.)

The first great step, then, in the carving out of the present scenery of the Highlands appears thus to have been taken by the sea. The hard, gnarled crystalline rocks that had been upraised by subterranean movements were planed down by the waves, and once more sank into the deep under a thick mass of their own ruins. In the next stage we find this levelled sea-bottom raised anew into land. It is the changes of that land-surface which remain to be traced.

FIG. 9.—SECTION OF THE GRAMPIANS FROM INVERNESS TO KIRRIEMUIR. (Vertical and horizontal scale the same.)

Inverness.

Spey.

Ben Muich Dhui.

Dee.

Glas Meal.

Cat Law.

CHAPTER VI.

ORIGIN OF THE HIGHLAND VALLEYS.

AT what geological epoch the re-elevation of the area of the Highlands above the sea took place has yet to be determined. It must have been very early, for the denudation which has been accomplished since then demands a vast interval for its completion. Such parts of the high grounds as did not sink beneath the sea-level during the marine denudation just described, would of course be exposed to constant sub-aërial waste at the same time that their shores were eaten away by the waves. And no sooner did the submerged land rise again into air, than it became in like manner a prey to the atmospheric agencies of erosion. Rain, falling on the new land, would find its way by the readiest paths to the ocean, and the various channels which it took would year after year be deepened and widened, both by the action of the running water and by springs and frosts. It is, I believe, by the working of these slow, silent, unobtrusive forces during the lapse of those vast geological ages which have passed away since the time of the Lower Old Red Sandstone, that the ancient table-land of the Highlands has been cut into the present system of mountains and glens.[1] In looking at the results of these changes we are very apt to regard them as a completed process—one that, after long ages of preparation and progress, has at last reached its end. But in sober truth the same process is going on still. Instead of coming in at its close, we look into its working when it has still, perhaps, far more to accomplish than has yet been done. With the exception of the ice of the glacial period, the same powers of waste are here still at work, and in watching their slow but certain progress we see exemplified before us the same kind of action which has brought mountain and glen to their present forms, and which may be destined in the long ages of the future to continue until it has worn down the solid land, and mountain and glen have alike disappeared.

It may not be uninstructive to look for a little at the progress of this waste, as it may be seen in many a Highland glen. A river, as has been already shown, digs out its ravines by gnawing its way backwards towards its source, and if we wish to study how this erosion takes place, we must visit the higher

[1] See Professor Ramsay's Lectures on the Physical Geology of Great Britain. Lect. II.

end of the gorge, and mark the action of the waterfalls and rapids. In like manner, a valley creeps backward and upward, and its origin and progress are sometimes best explained by what may be seen at its upper end. Valleys have step by step carved their way into the very heart of the high grounds; they may be watched even now eating deeply into the flanks of the mountains. And thus throughout the Highlands we meet everywhere with some rocky glen or rough mountain-slope,

> "Some tall crag
> That is the Eagle's birth-place, or some peak
> Familiar with forgotten years,—that shews
> Inscribed, as with the silence of the thought,
> Upon its bleak and visionary sides,
> The history of many a winter storm."[1]

Nowhere can a better illustration of this universal waste be found than among the deep glens and corries round the flanks of Ben Nevis. If the observer be sure of foot and steady of eye, let him ascend that mountain, not by the regular track, but up the long and almost equally lofty ridge which lies to the east, and thence along the narrow and somewhat perilous col which circles round to the southern front of the great Ben. The ascent lies first among heathery slopes, channelled with brooks of clear cold water, and roughened with grey, worn, and weathered hummocks of schist and granite. Blocks of granite

[1] Wordsworth, Excursion, B. i.

of every size cumber the ground, standing sometimes on rocky knolls, and sometimes half buried in morass. That the frosts of many a century have been busy here, is shown by the countless boulders and protruding knobs of rock which have been split open along their joints. Slanting up the mountain, the observer has leisure to remark, as he crosses streamlet after streamlet, that their channels, sometimes cut deeply into the solid rock, are evidently the work of the running water. He finds them grow fewer as he rises.[1] On the slopes, too, the boggy peat and shaggy heather begin to give way to long streams of angular granite blocks, among which the scanty vegetation is at last reduced to mere scattered patches of short grass and moss, with here and there a little Alpine plant. A wilderness of *débris* now covers the bald scalp of the mountain. The solid granite itself cannot be seen through the depth of its own accumulated fragments; but when the crest of the height is gained, the rock is found peering in shattered fragments from amidst the ruin. This narrow mountain ridge is then seen to rise between two profound glens. That to the north-east is crowned by a rampart-like range of pink-hued granite cliffs,

[1] The highest spring noted by Petermann on the flanks of Ben Nevis was at a height of 3,602 feet, or 766 feet below the summit. Another spring on Ben Aulder, one of the highest of the Grampians, was found by the same observer to be 3,650 feet above the sea. Edin. New. Phil. Jour. xlvii. 316.

from which long courses of *débris* descend to the bottom. The glen that lies far below on the south-west is overhung on its further side by the vast, rugged precipice of Ben Nevis, rising some fifteen hundred or two thousand feet above the stream that wanders through the gloom at its base. That dark wall of porphyry can now be seen from bottom to top, with its huge masses of rifted rock standing up like ample buttresses into the light, and its deep recesses and clefts, into which the summer sun never reaches, and where the winter snow never melts. The eye, travelling over cliff and crag, can mark everywhere the seams and scars dealt out in that long warfare with the elements, of which the whole mountain is so noble a memorial.

But, passing from the contemplation of the glens on either side and their encircling ramparts of rock, let the observer pick his way southward along the mountain of which he has now gained the top. He will soon find that, from a somewhat rounded and flattened ridge, it narrows into a mere knife-edged crest, shelving steeply into the glens on either side. It is sometimes less than a yard broad, and as it is formed of broken crags and piles of loose granite-blocks, it affords by no means an easy pathway. The rock here, as usual, is traversed with abundant joints. Of these the rains and frosts have made good use, and the result has been to shatter the summit of the ridge, and strew the slopes far below with its ruins. The process of waste may be seen in all its stages. In one part the solid granite is only beginning to show its lines of jointing, in another these lines have commenced to gape, giving the rock the appearance of rude uncemented masonry, and the severance can be traced on its way until the sundered mass has fallen over, and lies perilously poised on a ledge below the crest,

> "As if an infant's touch could urge
> Its headlong passage down the verge."

So narrow is the edge of the ridge in some places that a single block of granite may split into two parts, of which one would roll crashing down the steep slope into the valley on the left hand, while the other would leap to the bottom of the glen on the right. In this sharp form the ridge divides, one arm sweeping round the head of the glen on the north-east side, while the other circles westwards to the shoulders of Ben Nevis.

A more impressive lesson of the waste of a mountain-side, and the lowering of a mountain-top could hardly be found. The narrow ridge is a mass of ruin, like the shattered foundations of an ancient rampart, and its fragments have been thickly strewn on the steep declivities below. The larger pieces lie as a whole nearest the crest, though many a huge block has toppled down into the depths of the glens. When detached from the solid granite, they still remain a prey to the same ceaseless wear and tear. Rain, frosts, and snows split them up yet farther, and then, as they slowly tumble over each other in their downward course, they become by degrees smaller, though still retaining their dry, angular surface. At last, broken up many times in succession, they find their way down into the stream that threads the bottom of the glen. There, chafed against each other and the rocky channel, they are rolled into shingle and gravel, until at last, in the form of fine silvery sand, the waste of the far granite peaks is either spread out in the quiet reaches of the stream, as it winds through the valley, or hurried thence by the floods, and swept out into the waters of Loch Eil. The various agencies of erosion thus work steadily in concert; those that are wearing down the flanks of the mountains casting no more *débris* into the streams than these can in the end sweep away. Hence each glen is insensibly widened and deepened, and each mountain, as it decreases in circumference and in height, silently proclaims,

> "The memorial majesty of Time,
> Impersonated in its calm decay."

Watching this ceaseless excavation, we cannot fail to be powerfully impressed with the efficacy of these silent atmospheric agencies of waste. Nor can we well hesitate to admit that, during a period of sufficient vastness, these agencies could have carved out each of the glens without the aid of subterranean convulsions. The only element to be granted in the problem is time—a concession which the progress of geology is every year making more imperative and exacting. In the scenery just described, there is no evidence of any open fissures or dislocations to which the glens could have owed their origin. On the contrary, these deep, narrow valleys are crossed at their upper ends by the precipitous junction of their sides, and the connecting wall of granite, though seamed and cracked by the weather, shows no trace of any great subterranean fracture. It is indeed quite possible that there may be many faults in this as in the other mountainous parts of Scotland, and, as will be afterwards shown, these faults may have contributed at first to determine the lines of some of the glens. But the profound concavity of these valleys cannot possibly, I think, arise from the sundering of the sides of a fissure. The two deep glens to the east of Ben Nevis are separated by the narrow granite ridge already described; if one of them had been opened by a rending of the mountain, the expanding of its sides would either have effectually prevented the other from being similarly rent, or would have closed it up had it been already formed. And if it be contended, in answer, that the two narrow open chasms might have been produced by underground movements, and afterwards enlarged by running water and the other powers of erosion, such an hypothesis would really assign most of the work to these gentle agencies. Not only so, it might necessitate the admission that every Highland mountain from which glens diverge is an independent focus of disturbance—an idea wholly disproved by the geological structure of the Highlands. The radical error of such an explanation lies in the fact that it assumes the present surface of the country to be, approximately at least, the same as that which witnessed the supposed dislocations to which the glens are attributed. It thus ignores the vast denudation which the whole of that region has undergone. But, as I have already tried to explain, hundreds and thousands of feet of hard rock have been worn away from the tops of the Highland mountains since the rocks were metamorphosed, contorted, and faulted. The existing outline of the country, therefore, must represent the results, not of these ancient subterranean movements, but of a long subsequent process of denudation. And, in looking at the disposition of the Highland glens and straths, their winding course, their orderly system, and their complete subservience to the drainage of the country, it seems hard to see how any one can call in question the conclusion that the valleys have been made by sub-aërial waste.

Valleys crossing Watersheds; Passes or Beallochs.

There is one other useful lesson which may be learnt on the sides of Ben Nevis. The upper ends of the glens just described are formed by an abrupt wall, or steep slope of granite, with a narrow, knife-edged summit. On the other side of this rampart is the head of another glen, trending in the opposite direction, so that the granite ridge is a kind of partition-wall between two glens running northward, and one running southward. Now it is easy to see that as this wall is every year insensibly decreasing in height and thickness, the time will come when it will disappear. The glens, eating their way backward towards each other, have reduced the space between them to narrow limits. This intervening and lessening space is doomed in the end to be wholly removed, and then one long glen will run along the east side of Ben Nevis, with perhaps a low, scarcely perceptible watershed marking where the narrow transverse granite-ridge now rises. Such seems to have been essentially[1] the origin of the

[1] It is of course to be borne in mind, that the present forms of the Highland glens show the action of the ice of the glacial period, and that, though essentially the work of atmospheric waste and streams, the erosion of these hollows has been greatly helped by glaciers, as will be shown in the sequel.

numerous long Highland valleys in which there is a central inconspicuous point whence the water flows in opposite directions. The sea and other powers of degradation may have lent their aid in completing the levelling of the barrier between the advancing glens. But the first, and perhaps the main part of the work was probably done by running water and frost.

Examples of this form of scenery will readily occur to any one who has paid even a cursory visit to the Highlands. Most of the high roads are carried through such continuous valleys, where the watershed is often so imperceptible that it may be crossed unawares, even by one who is on the outlook for it. The road from Loch Carron, across Rosshire to Contin, seems to run along one great transverse valley bounded on either side by lofty hills, yet if the tourist watches the flow of the water, where the road reaches a height of somewhere about a thousand feet above the sea, he will observe that the streams flow off in opposite directions, one set turning eastwards, and falling into the Cromarty Firth, the other bending south-westward, and joining the Atlantic in Loch Carron. There is no mountain, hill, or ridge, not even a marked mound, to make the watershed; the valley of Loch Carron ascends inland, and imperceptibly merges into another valley, which descends to the sea on the opposite side of the island. In like manner the post-road from Arisaig to Fortwilliam passes through a long valley which connects the head of Loch Aylòrt with the head of Loch Eil. Another transverse glen runs from the banks of Loch Fyne, across Cowal to the head of the Holy Loch, and contains the long, narrow fresh-water lake known as Loch Eck. The great north road from Perth to Inverness runs up the valley of the Garry, and thence across the watershed of the Grampians into Glen Truim, the connexion between the two valleys being made by the Pass of Drumouchter—a wild glen in which, among the rubbish-mounds of old glaciers, the water flows partly northward down Glen Truim, and partly southward down Glen Garry. In these, and innumerable other instances, the two glens seem to have been deepened and cut backward toward each other by the action of sub-aërial waste, sometimes aided, perhaps, by the sea, until they have at last united—giving rise in this way to long valleys, that run across high hill-ranges.

Sea-lochs.

There is another kind of valley in the Scottish Highlands deserving of special notice. To a small extent on the east coast, but on a great scale along the western side of the island, the sea runs inland in long, narrow firths or fiords. Each of these sea-lochs terminates at the mouth of a glen or strath, and receives there the collected drainage of the interior.

There is commonly no marked line of demarcation between the land-valley, watered by its brawling brook or river, and the sea-valley, filled with the ebbing and flowing tides. The grassy slopes or rocky declivities which form the sides of the one, run on to form the sides of the other. If we could depress the land below the present sea-level, and send the salt water far up into these inland glens, they would become fiords, comparable in every respect to those at present indenting the western coast. If, on the other hand, we were able to upraise the land, the sea-lochs, emptied of their salt water, would become land valleys, and it might in the end be impossible to say where the former limit of the tides had been. In each case we see what is in truth one valley, part of it being submerged and part open to the sky.[1] It is a common idea that the indented character of the western coast is due to the unequal encroachments of the sea. But even a superficial acquaintance with the usual features of these sea-lochs ought to disprove such a notion. For it is well known that the inlets are, as a rule, deep, sometimes much deeper than the sea outside. But the sea, as we have seen, cannot scoop out deep hollows; its erosive power is confined to its upper part, and its tendency is to plane down its shores. Instead of excavating a fiord, it can only

[1] See Professor Ramsay's remarks, to the same effect, in his paper on the origin of lakes. Quart. Jour. Geo. Soc. xviii. p. 185.

level the rocks that are lashed by its breakers. If we are right in regarding the land-valley as the result of an erosion performed by running water and the other sub-aërial denuding agents, a similar origin must be assigned to the continuation of the same valley under the sea-level. And that such is the true explanation I have no doubt. The sea-lochs of the west coast are thus not inlets cut out by the waves, but old glens that have been submerged beneath the sea. They prove that at least the west side of the island has sunk down under the Atlantic in a comparatively recent geological period, and that the tides now ebb and flow where of old there was the murmur of brooks and waterfalls. This conclusion will be seen to possess much interest in its relation to the history of the glacial period in Scotland.

If this view of the origin of the western sea-lochs be correct, it is natural to expect that traces of different stages of the submergence should be found; that, as the downward movement of the land went on, some lake-basins in the valleys should have been carried far down beneath the deep, that others at a higher level should have sunk but a short way below the waves when the depression came to an end, and that others should have no more than escaped when they had approached within a few yards of the sea-level. Examples of these various steps in the process will occur to those who are familiar with the western coasts. Of the first, Loch Fyne is a notable illustration, as it deepens a little south of Tarbert into a basin 624 feet below the surface of the loch, and shallows northward and southward. If Loch Fyne were now a land-valley this depression would be filled with a lake. Of the second stage, Loch Etive forms a good example. That fiord narrows at Connal Ferry, and across the straitened part runs a reef of rocks covered at high water, but partly exposed at ebb. Over this barrier the flowing tide rushes into the loch, and the ebbing tide rushes out, with a rapidity which, during part of the time, breaks into a roar of angry foam like that of a cataract. The greatest depth of the loch above these falls is 420 feet; at the falls themselves there is a depth of only six feet at low water, and outside this barrier the soundings reach, at a distance of two miles, 168 feet. Loch Etive is thus a characteristic rock-basin, and an elevation of the land to the extent of only twenty feet would isolate the loch from the sea, and turn it into a long, winding, deep fresh-water lake. Of the third stage, where the lake has been brought down close to, but has not quite reached, the present sea-level, Loch Maree, Loch Morar, and Loch Lomond may be taken as illustrations.[1] If the downward movement were to

---

[1] For the sake of simplicity I have left out of account the later upheavals of the land, which have to a trifling extent restored what the older depression had submerged. Maritime lakes, such as Loch

recommence, these lakes would ere long be turned into arms of the sea.

It is curious to watch the efforts made by the land for the recovery of its lost territory. At the head of many of the sea-lochs, as for instance at Loch Carron, the rivers are pushing their alluvial flats out into the salt water, and gradually driving it backward, regaining in this way, step by step, the site over which they once rolled. In other cases the tides and currents of the sea itself are raising barriers against it. Thus Loch Fyne is nearly cut in two by the long sand-bar thrown up by the tides at Otter. The Gareloch in like manner is almost barred across by the similar spit which runs out at Row. In each instance a powerful tidal current tends to keep the narrow channel open by sweeping fresh accumulations of sediment out to sea.

The eastern seaboard of the Highlands presents a striking contrast to that on the west side of the island. Instead of winding far into lochs and kyles, or beating round peninsulas and islands, the sea there rolls along a coast-line that runs mile after mile in a persistent course, interrupted merely by trifling indentations. It is only in the angle between the shores of Caithness and those of Elgin that inlets

Lomond, must of course have been arms of the sea when the land stood at the level of the 40-foot raised beach, that is, forty feet lower than it stands to-day.

occur in any way comparable to those of the west. Beginning at Loch Fleet, on the sea margin of Sutherlandshire, we pass in southward succession the Firths of Dornoch, Cromarty, Beauly, Inverness and Moray. These resemble in outline some of the narrow fiords of the west, Cromarty Firth approaching nearest to the western type. But they show many essential points of difference—their shores, as a rule, are low and are usually formed, not of rocks shelving down below the water, but of raised beaches and slopes of boulder clay. They have thus a smoothness and tameness of character which is wanting in the western sea-lochs. Moreover, they are, as a whole, shallow, the tide leaving wide flats at low water, and sometimes even forsaking the firth altogether. At Loch Fleet, for instance, advantage has been taken of this feature, and a strong mound having been built across the firth to keep back the salt water, the higher part of the loch has been turned into dry land. Nearly the whole of these firths are conspicuous for the sand-bars at their mouth which have been thrown up by the tides above high-water mark, and run out from either bank, striving as it were to form a completed barrier against the sea outside. The mouth of the Inverness Firth is a notable example. From the east side a long low sand-ridge stretches half way across and bears Fort George at its extremity. From Fortrose on the west side another spit of the same kind runs out into the middle of the firth, the two bars actually overlapping each other, so that if they lay exactly opposite they would meet and turn the Inverness Firth into a lake. The Dornoch Firth is likewise intersected by a long ridge of gravel extending from the south bank at Meikle Ferry, and with a corresponding but smaller spit a little further down on the north bank.

The Cromarty Firth perhaps furnishes to the geologist more matter of interesting inquiry than any of the others. At its upper end, where the Contin brings down the drainage of a large tract of eastern Ross-shire, the firth has been so encroached upon by the advance of the river alluvium, that several square miles of sandy and muddy flats are laid bare at low water. From these higher shallows the firth stretches, with a tolerably uniform breadth of rather more than a mile, as far as Invergordon, where it reaches a maximum depth of seventy feet. It then expands into a wider basin, forming the sheltered spacious anchorage for which the inlet is so well known. This expansion of the firth is not so deep as the narrow channel at Invergordon. But a little to the east of the town of Cromarty, where it suddenly contracts to a breadth of less than a mile, it shelves down to a depth of 170 feet, and passing between the two precipitous headlands of the Sutors, enters the open

Moray Firth. One who approaches from the east is at once struck with the narrow chasm-like entrance of the Cromarty Firth cut through a long lofty range of red sandstone precipice. It is wholly unlike the mouth of any other firth in the country, for it is not the seaward expansion of a land valley, but seems, in some abnormal fashion, to have been broken through a high barrier of hard rock. And, in actual fact, it is an abnormal opening, and not the original mouth of the firth.

To understand the present and the former conditions of this sea-loch, one should ascend the heights behind the town of Cromarty. Looking from the crest of the Black Isle over a scene which the pen and hammer of Hugh Miller have made classic ground to the geologist, he sees below him the firth filling the ample basin of Cromarty, creeping over sandy flats in the bay of Nigg, and turning thence abruptly to the south-east to force its way between the Sutors. The north-west side of the estuary is formed by a gently-sloping declivity stretching towards Tain. The ridge of the Black Isle rises on the south-east side and runs north-eastward through the two Sutors into the long promontory of Tarbat Ness. The north-easterly line of the firth is continued from the Bay of Nigg to the Dornoch Firth by the low valley or plain of Easter Ross, bounded on the one side by the rising grounds between Invergordon and

Tain, and on the other by the ridge of the Black Isle and its prolongation beyond the North Sutor. Even from a distance it is not difficult to see that this low valley must have been the original outflow of the firth. From the heights above Cromarty the eye looks along the northward drift-covered slope of the Black Isle, and if it were not known that the sea flows between the Sutors, it might readily be supposed that the ridge extends in one long continuous and unbroken line from the head of the firth away to Tarbat Ness. It could not be suspected that this ridge is actually cut through by the present narrow precipitous opening. If, then, the Cromarty Firth once entered the Firth of Dornoch between Tarbat Ness and Tain, how has it come to join the Moray Firth by so abrupt and narrow an outlet as that which is bounded by the Sutors?

Here, perhaps, if anywhere, the believers in the efficacy of underground convulsions might make a bold stand. How could the firth, they might ask, quit its old channel and take one so widely different, unless the new outlet had been opened for it by an earthquake shock? The two Sutors seem to

> "Stand aloof, the scars remaining,
> Like cliffs which have been rent asunder,
> A dreary sea now flows between."[1]

It may be impossible absolutely to prove the origin

[1] Coleridge's *Christabel*.

of the deep gorge between the Cromarty Sutors; but it seems to me that there is evidence in the neighbourhood strongly in favour of the supposition that the work has been mainly done by running water. Let the observer cross to the Nigg side of the firth, and traverse the high ridge which is prolonged from the Black Isle. The rocks there are well *ice-worn*, the moulded knolls and the striations upon them running in an E.N.E. direction. Passing over to the eastern coast, a line of lofty red sandstone cliffs—the continuation of those on the south side of the entrance to the Cromarty Firth—rises from the rocky beach and extends northwards to Shandwick. About a mile south of that fishing hamlet, a depression or valley, running across the ridge, descends into the flat plain of Easter Ross on the west side, and to the edge of the sea-cliff on the east. The bottom of the valley is coated with boulder clay, through which two runnels, diverging from the low watershed of the hollow, flow in opposite directions. One of them trickles westward into the low grounds between the Bay of Nigg and the Dornoch Firth, the other having a steeper and shorter declivity has cut its way deep into the red sandstone, forming there a ravine which breaks through the cliff and descends upon the beach. The streams are thus busy digging out the valley and working their way backward to each other. The same process may have been going on previous to the Age of Ice, for the valley is older than the boulder clay. Nor can it be doubted, that if time enough be allowed, these tiny rivulets will in the end cut down the ridge, and allow the sea to force its way to the plains of Easter Ross through a narrow gorge similar to that between the Sutors. It will be observed that this form of denudation is the same as that which has been already referred to as the probable origin of mountain passes, and of long nearly level valleys crossing watersheds. Only, in the present instance, the changes are going on close to the sea level, and it can be readily seen how much, in the latter stages of the process, may have been done by the co-operation of the wild breakers of the Northern Sea. As I have just said, it cannot be demonstrated that the present outlet of the Cromarty Firth has been opened in this way, but I think no one can watch the effects of the streams near Shandwick without a conviction that the explanation has high probability in its favour.[1]

The greater depth, steepness and number of the fiords on the west side of the island, may be partly attributed to the greater height of the country on that side.

[1] If, as is probable, this excavation was begun long before the glacial period commenced, it may have been greatly helped by the mass of ice which filled up the Cromarty Firth, and, as shown by the striations, slanted across the ridge of the Black Isle. There could not fail to be a great pressure of ice through the gorge between the Sutors.

Perhaps also a more extensive submergence of the West Highlands may have helped to indent the western sea-board, by allowing the waters of the Atlantic to run far up among the recesses of the mountains, and thus fill the glens.[1] The east coast, as a whole, is lower and farther from the hills, and its rivers enter the sea, not between the steep sides of rocky glens, but among level or gently undulating lowlands. A submergence of the seaward ends of these river valleys would not make the coast-line resemble the coast-line of the other side of the kingdom, save only in such parts as the east of Sutherland, where the mountains and their narrow glens come within a short way of the sea margin. Long narrow sea-lochs, like those of Inverness and Argyle, are thus almost wholly absent from the eastern shores; where they do occur, it is just where the character of the neighbouring ground approaches most to that of the Atlantic sea-board. They are crowded together in the angle between the counties of Caithness and Elgin, where the mountains of Sutherland, Ross, and Inverness, come closest down upon the North Sea. At the same time, there may be some other cause, not at first apparent, to which the singular difference between the outline of the east and west coast may

---

[1] Does this suggestion derive any support from the rise of level of the deposits of Arctic shells traced from the Firth of Clyde into Lanarkshire?

be due. For it must not be forgotten that precisely the same kind of diversity, but on a larger scale, is prolonged into the Scandinavian peninsula. The west side of Norway, with its thousands of deep fiords and inlets, is a magnified piece of west Highland scenery, and the undulating drift-covered plains of Sweden, with their comparatively unindented coast-line, along the Gulf of Bothnia, recall those of the east side of Scotland.[1]

VALLEY SYSTEMS OF THE HIGHLANDS.

Having followed in some detail the operation of atmospheric waste in wearing hollows on the surface of the land, and giving rise to capacious valleys, we may now look at the general results of this process upon the country at large, and try to trace the probable origin of the various valleys and river basins. To carry out this inquiry thoroughly, would demand larger and better maps, alike topographical and geological, than yet exist, as well as a far more minute and extended series of observations than has hitherto been made. I cannot pretend to do more than offer a mere outline of the subject so far as I have been able to apprehend it.

[1] There can be no doubt that at least some of the later touches were given to the scenery both of Scandinavia and of Britain by the agencies of the glacial period.

If the reader will take a map of Scotland, and consider with attention the arrangement of the glens and straths of the Highlands, he will remark, I think, that two systems of valleys can there be made out. Of these, the one has a general trend from north-east to south-west, the other from north-west to south-east, the one series being thus transverse to the other. Between these two systems there are many intermediate examples; nay, the same valley may sometimes incline to the one and sometimes to the other. But even with such doubtful forms, the two main systems remain tolerably persistent. As illustrations of the north-east and south-west, or *longitudinal* valleys, reference may be made to the Great Glen, Lochs Carron, Shiel, Linnhe, Fyne, Awe, and to the valleys of the Findhorn, Spey, and Loch Tay. The north-west and south-east, or *transverse* valleys, are conspicuously seen on the eastern and western sides of Sutherland and Ross, in Inverness and Argyle, and throughout the eastern half of the kingdom, from the headlands of Aberdeenshire to the Forth.

It must be granted that no clue remains as to what may have been in detail the original form of the surface, when the metamorphosed corrugated schist, gneiss, and quartz-rock of the Highlands were first upheaved above the sea. The land possibly rose in one or more broad ridges, having a north-easterly direction with a seaward slope along either side.

Rain falling on such a surface, as has been already remarked, would of course run off as rapidly as it could, selecting for itself the readiest paths which would naturally lie down the declivity from the summit level of the ground to the sea. The drainage would thus collect into streams flowing at right angles to the trend of the land, one series descending to the south-east, and the other to the north-west. When these streams once began to flow they would tend to keep the channels they had chosen. The sides of these channels would simultaneously be attacked by rain and springs, and in due season by frost, so that the ancient Highland table-land, instead of being cut into a network of deep chasms, such as would have resulted from the action of the streams alone, became a prey to a much more diffused and general waste, by which the water-courses were gradually widened and deepened into valleys. The detritus of the slopes was washed down by rain into the streams which both carried off this detritus, and continued to excavate their own beds. Thus valleys slowly sank, as it were, into the table-land. To this process I am inclined to attribute the origin of those *transverse* Highland glens and straths, which run more or less directly across the strike of the rocks, and descend to left and right of the main chain of high grounds.

The excavation of these valleys formed part of the

general denudation and lowering of the surface of the country. But the rain, springs, frosts and streams, which were busy everywhere, would have their effects largely determined by the varying nature of the rocks. As the prevailing strike of the strata forming the Highlands passes from south-west to north-east, the ridges of harder and softer material run in that direction. The larger faults, the anticlinal and synclinal axes, follow the same line. It would be natural, therefore, that choosing some original depressions or hollowing them along bands of more easily worn rock, the powers of waste, besides deepening and widening the transverse water-courses, should also in the end cut out a series of long parallel valleys corresponding to the strike of the rocks—that is, from south-west to north-east. Such seems to me to be the probable origin of the *longitudinal* valleys, such as the Great Glen, Upper Loch Fyne, Loch Tay, and the numerous parallel glens and fiords of the south-western Highlands. It must be remembered, however, that even these valleys do not in every case, nor even, perhaps, in any case, rigidly conform to the line of strike or fault or axis which at first fixed their direction. The original determining cause has not been permanent or powerful enough to prevent the eroding agents from cutting deeply into the rocks on either side, and turning a straight valley into a more or less winding one. Still, no one can fail to remark that these north-east depressions are, as a whole, far more straight and parallel than those which run in a transverse direction. If the explanation of their origin here given is correct, this difference may be thus accounted for. The direction of the valleys that diverge to right and left from the main line of water parting was determined by the first seaward flow of the rain across the strike of the rocks of which the newly upraised land was formed. There were no great geological lines to guide the drainage which thus found its way among the irregularities of the sea-worn slopes of the ancient table-land. Hence the valleys, while they preserve a general trend at right angles to the main chain of heights, are found to sweep into broad curves, cutting across many different rocks in succession, and sometimes coming back again upon themselves, in a manner for which their geological structure will not in anywise account. The valleys that follow, on the other hand, a south-westerly and north-easterly course, run in a general sense parallel to the strike of the rocks, and their straightness and parallelism have no doubt been determined by the trend of the rocks along the edges of which they have been worn.[1]

[1] Although I have long held the belief of Hutton, that our valleys are mainly the work of atmospheric waste, the history of the process of their excavation was but dimly understood by me until the appearance of the admirable paper by my colleague, Mr. J. B. Jukes, on the River-Valleys of the South of Ireland. The explanation above given of

The watershed of the country runs southward from Cape Wrath to the head of Loch Quoich, whence it turns sharply eastward between Lochs Lochy and Oich to the Monadhleadh Mountains and the hills above the head of Loch Laggan. It then follows a curving southerly course, past the west end of the Moor of Rannoch, and the Brae Lyon mountains to Crianlarich, thence across Ben Lomond, and south-eastward over the Campsie Fells into the wide Lowland valley. Skirting the south-western parts of Linlithgow and Midlothian, it strikes up into the Pentland Hills, thence due south between the valleys of the Clyde and Tweed to the Hatfell heights, from which it sweeps across the southern uplands to the Cheviot Hills. To the west or left side of this line the water flows into the Atlantic; to the east or right side it enters the German Ocean. This watershed, it will be seen, runs irregularly across the course of the main hill ranges, just as the watershed of Europe, for example, from Gibraltar to St. Petersburg, sweeps in a widely curving line athwart the chief mountain chains. Owing to the comparative

of the probable relative history of the Highland valleys is essentially the same as that which he has proposed for those of Ireland, see *Quart. Jour. Geol. Soc.* xviii. (1862) p. 378. There must still be many difficulties in the application of these views, arising, among other causes, from a want of minute and correct topographical geology and good maps. But they explain so much and reconcile so many discrepancies, that I have gladly availed myself of them in trying to gain some insight into the history of Highland scenery.

steepness of the west side of the island, the line of water-parting keeps much closer to the Atlantic than to the German Ocean. Hence by far the larger area of the country is drained into the latter sea. In the northern or Highland half of Scotland no river of any notable size enters the Atlantic, while on the east side, the Spey, the Dee, the Don, the Tay, and a number of smaller, but still considerable, rivers carry the drainage of the mountains to the sea.

On the western side of the watershed, as it runs down Sutherland, Ross, and Inverness-shire from Cape Wrath to Loch Quoich, almost every great valley that enters the sea comes down from the south-east and has its seaward portion filled by the tides of the Atlantic. The only marked exceptions are Lochs Keeshorn, Carron and Alsh. These belong to the longitudinal or north-easterly system of valleys. In Loch Torridon there is an approximation to a union of the two series; for the long valley that comes across from Kinlochewe enters Upper Loch Torridon from the north-east, and not until the mouth of that loch is reached does the fiord turn into the prevailing north-westerly course. So at the head of Loch Duich the dark Alpine defile of Glen Shiel opens from the south-east, while another glen comes down from the Bealloch of Kintail on the north-east, and the two valleys unite at a right angle to form Loch Duich.

The rest of the Highlands to the west of the watershed exhibits a predominance of the longitudinal valleys, the chief being Loch Linnhe, Loch Awe, Loch Fyne, Loch Long and the glens of which they are seaward continuations. In no part of the Highlands can the direct effect of geological structure in determining the trend of ridges and hollows be better seen than on the east side of the Sound of Jura. The long narrow sea-lochs of Craignish, Swene, Killisport and Tarbert have been cut out along bands of schist and slate, which all run from south-west to north-east; and even the direction of the little creeks and headlands, and the form of the islands have been largely determined by the same cause. This is the reason why along that part of the coast, island, promontory, bay, and fiord seem all ranged in parallel lines bearing towards the south-west.

Throughout the greater part of the eastern side of the country the water, instead of finding its way to the sea by hundreds of independent streams, is gathered into wide basins, and enters the German Ocean in large rivers. From John o' Groat's House to the Dornoch Firth the valleys belong to the transverse or south-east group, like those on the opposite side of the watershed. From the Dornoch Firth eastward to the borders of Aberdeenshire, the valleys of the Cromarty Firth and the Beauly, Ness, Nairn, Findhorn and Spey are illustrations of the longitudinal series.[1] Some of these streams, as the Findhorn and Spey, show well how, though the general direction of the valley has been marked out by the strike of the rocks, the river has not scrupled to cut its way to right and left across the strike, and now wanders in wide winding curves, though its course as a whole is parallel with the trend of the strata. On the eastern side, from Buchan to the Firth of Tay, the transverse system prevails. It is shown by the valleys of the Don, and the Dee, the two Esks, the Isla, the Shee and Ericht, the Garry and Tay.

Let us look for a little at the basin of the Tay, where the two groups are characteristically developed. The river drains an area of 2,750 square miles, and pours a larger amount of water into the ocean than any river in Britain.[2] It rises among the lofty mountains between Strath Fillan and the head of Loch Fyne, and after a short course down Strath Fillan, strikes in a long north-easterly line down Glen Dochert into Loch Tay, and thence to a point below Grandtully Castle, where it turns sharply round into the transverse valley of the Tummel. I have already had occasion to point out that Loch Tay runs along an

[1] Why the longitudinal valleys should have the ascendancy over the transverse ones is a question not easily answered. Unequal upheaval or depression of the country may have had a powerful influence upon the river-systems.

[2] The discharge is estimated 273,000 cubic feet per minute. See A. K. Johnston's *General Gazetteer*, edit. 1864.

anticlinal axis.[1] The whole of this part of the valley of the Tay, from the head of Glen Dochert to the Tummel, corresponds with the general strike of the rocks, and no doubt owes its origin and direction to that coincidence. It is one of the longitudinal valleys of which the erosion has been determined by geological structure. But at its north-eastern end it is abruptly cut off by the transverse valley which comes from the north-west down Glen Garry. A few miles further north the same transverse valley truncates in like manner the strath of the Tummel, and thus the drainage, which had flowed for many miles through the mountains towards the north-east, is turned back at a right angle to its previous course, and carried to the south-east, out of the Highlands into the wide plains of Strathmore.

If one may venture to offer a possible explanation of the history of these valleys, I may suggest that the course of the Garry was defined before any marked hollows ran transverse to it. This valley comes down from the heart of the mountains and passes away to the south-east, like those of the neighbouring Forfarshire rivers. It thus follows what would be at first the natural descent of the water, by the shortest and readiest route from the high grounds

---

[1] The surface of Loch Tay is 355 feet above the sea, and its depth varies up to 600 feet, (A. K. Johnson, *loc. cit.*). Its bottom is thus nearly 250 feet below the level of the sea.

to the sea. The long valley of Loch Tay and Glen Dochert, on the other hand, has been excavated along an anticlinal axis or fold of the quartz-rocks and schists,[1] by the same atmospheric agencies which have carved out Glen Garry. But the latter glen had already been deepened sufficiently to carry off all the drainage that might come from any longitudinal valleys. These valleys, therefore, could not cross it, but must needs pour their waters into it, and help in this way to increase the rapidity of its excavation.[2] To the south-west of the valley of the Garry, Tummel, and Tay, there is a number of short valleys or depressions running parallel to it across the ridge of hills between Loch Tay and the Lowland border. Some of these, like Glen Ogle, pass completely across this ridge, and on the north-west bank of the Loch Tay Valley they find their counterparts in depressions which trend to the north-west towards Rannoch. It almost seems as if these depressions once ran south-eastward, across what is now the deep hollow of Glen Dochert and Loch Tay, that hollow

---

[1] The coincidence of a valley with an anticlinal axis may perhaps be traceable to an actual fracture of the strata along this line of severe tension. Not that the present sides of the valley are the sides of the fracture, nor even that there was ever an open fissure at the surface at all, but that after the removal of a great mass of rock by the sea, and other denuding agencies, the crack still gave rise to a feature above ground and guided the subaërial forces in their work of erosion.

[2] Let me again refer the reader to Mr. Juke's paper, already quoted, on the River-valleys of the South of Ireland.

having since then been gradually cut out so as to sever these transverse valleys and divert their drainage in great part into its own channel.[1]

[1] The erosion of the glacial period must not be forgotten. Loch Tay, like the other rock-inclosed Highland lakes, has had its basin scooped out, I believe, by land-ice; but the valley itself was probably there before the ice filled it. This subject will be discussed a few pages further on. With regard to the power of streams to shift their own course and cut their way to each other across a low watershed, it may be mentioned here in passing that the main watershed of the country between the Tweed and Clyde, crosses at one part a low valley through which it would be easy to cut a channel for the Clyde. Indeed, if good care were not taken of its banks, the Clyde would ere long dig the channel for itself and flow into the Tweed. In offering the above sketch of the origin of the Highland valleys, I make it chiefly as a suggestion in a subject full of difficulty, and standing much in need of thorough discussion. Though I am fully persuaded that these valleys are to be looked upon as the results not of subterranean movements, but of subaërial denudation, I have still very much to learn as to the way in which the process of excavation was carried on.

CHAPTER VII.

INFLUENCE OF ANCIENT GLACIERS AND ICEBERGS ON HIGHLAND SCENERY.

FROM the foregoing narrative it appears that a wide sea-worn tract of land has, since the time of the Lower Old Red Sandstone, been channelled out by running water, and general atmospheric waste, into systems of valleys, between which the intervening ground, having suffered less denudation, rises up into ranges of hills and mountains—the whole forming the framework of the Highlands of Scotland. We have now to trace the impress of another agency—that of glacier-ice and icebergs, no longer at work in our country, but which once distinguished the British Isles as markedly as the Arctic ice-fields characterise Greenland, and which played an important part in the changes that brought about the present scenery of the country. It is now well ascertained that during a comparatively recent geological period, the climate of the northern hemisphere was much colder than at present, and that in the British Islands, as well as in other countries where glaciers are now unknown, the land was enveloped in snow and ice. This part of the geological record is known as the Glacial Period. There is a growing belief among geologists that it was not an abnormal condition of things, but that in the past history of the globe there have been older cold periods succeeding each other after wide intervals.

In following the track of the ancient Scottish glaciers and icebergs, and noting how much they have influenced the scenery of the country, it must be borne in mind that the present great leading features of mountain and valley had been fixed before the ice-age began. The Highland hills rose then above the glens and valleys as they still do, and the glens and valleys wound in the same tortuous course towards the sea. The minor outlines of the surface, however, were, perhaps, in many respects, unlike those which it wears at the present day. There may have been, and probably there was, far more angularity and ruggedness about the contour of that ancient land. The passing of the long glacial period did much to remove such irregularities and smooth the general surface of the country. But though the ice must have worn down the valleys, it did not make them. They are to be attributed, I believe, to that earlier process of sub-aërial denudation which has just been sketched. Keeping in recollection, therefore, that hill and valley were grouped into their present arrangement before the ice began to settle down upon them, let us look for a little at the evidence from which this strange chapter in the country's history is deciphered.[1]

Track of the First Great Ice-sheet of the Glacial Period.

The surface of Scotland, like that of Ireland and large tracts of England, as well as the whole of Scandinavia and northern Europe, has a peculiar contour, visible almost everywhere, irrespective of the nature of the rock on which it shows itself, and therefore to be regarded as the result of one great process, acting upon all the rocks alike, long after they had been formed. This contour consists in a rounding and smoothing of the hills and valleys into long flowing outlines. What were once prominent crags have been ground down into undulating or pillow-shaped knolls, and deep hollows or gentler depressions have been worn in the solid rock, not at random, but in a recognisable system. These features may be seen throughout the whole of the country. At present we are concerned only with their development in the Highlands. It may seem paradoxical to speak of the well-known rugged Highland mountains as showing traces of a general smoothing of their surface. But such is really the case. There may be places, indeed, where from

[1] See Memoir on the Glacial Drift of Scotland; and *North British Review*, Vol. xxxix. p. 286, *et seq.*

height, or steepness, or some other cause, the smoothed surface was never communicated; and there is everywhere a constantly progressing destruction of that peculiar outline; the rains, springs, and frosts are re-asserting their sway, and carving anew upon the country its ancient ruggedness. Nevertheless, to an eye which has learnt to distinguish the characteristic flowing lines, there are not many landscapes in the kingdom where they cannot be traced. Even in the wildest Highland scenery, where the casual tourist may see nothing but thunder-riven crags and precipices, and glens blocked up with their ruins—

> "Precipitous black, jagged rocks,
> For ever shattered and the same for ever,"—

the geologist can often detect traces of the same universal smoothing and moulding. Nay, it is precisely amid such scenes that he is most vividly impressed with the fact that the surface of the country has been ground down by a vast general agent, for he there sees what are the natural outlines which the rocks assume when left to the ordinary attacks of the elements. Instead of smooth undulating outlines he notes craggy precipices and scars, here and there red and fresh, where the last winter's frosts have let loose masses of rock into the valleys below. He can trace how in this way, the hand of Nature is once more roughening the landscape, restoring to the hard rocks their cliffs and ravines, and to each knoll and crag a renewal of its former angularity. Yet his eye rests continually upon little bosses of rock, or even upon whole hill sides where, owing to a covering of drift or soil or to the nature of the rock, the change has gone on but slowly, and where he can still view the uneffaced traces of that wonderful process by which the surface of the country from Cape Wrath to the Solway has been worn and smoothed.

This wide-spread abrasion, however, is not only a general moulding of the country on the great scale. It can be traced on hills and crags of every size down to mere hummocks and knolls; nay, even to the merest knobs and protuberances; in short, not only is the general configuration of the surface affected by it, but it may be followed out upon all the little dimples and prominences on a freshly exposed surface of rock. The hardest rocks usually show its effects best, and when the soil and superficial detritus are stripped from them, their faces may often be seen to be as smoothly dressed as if they had been cut in a mason's yard, and were meant to form part of the polished ashlar-work of a great building. Further, not only are they thus planed down, they are traversed by long and more or less parallel ruts and striæ, varying in depth and width from mere streaks, such as might be made by a grain of sand, up to grooves like those worn in old pavements by the cart-wheels of successive generations. The fine striations may be seen descending into the hollows and mounting over the prominences of a rock, keeping all the while their general direction, with about as much regularity and persistence as they do over the most even surface. It is plain that in whatever manner these markings were produced, they must be due to no violent agent rushing like a *débâcle* across the country. They can only have been made in a quiet, leisurely way, by some force that paid little or no regard to the minor inequalities of the ground, but passed on with a steady persistent march, pressing grains of sand, pebbles, and even large blocks of stone upon the rocks below, in such a way as to leave there at last a smooth polished surface, marked by striations of varying coarseness, according as the rude polishing paste of detritus consisted of fine sand, or gravel, or boulders.

Now, just as the whole country has been smoothed, so has it at the same time, by the same agent, been striated. It is hardly possible anywhere to peel off the upper covering of clay and soil without laying bare a striated surface of rock, if the rock be at all of a kind to receive and preserve such markings. Moreover, these striations are distributed with a remarkable symmetry. They radiate from the main mountain masses outwards to the sea (see Map,

MAP OF SCOTLAND

The arrows indicate the movement of the land-ice of the Glacial period.

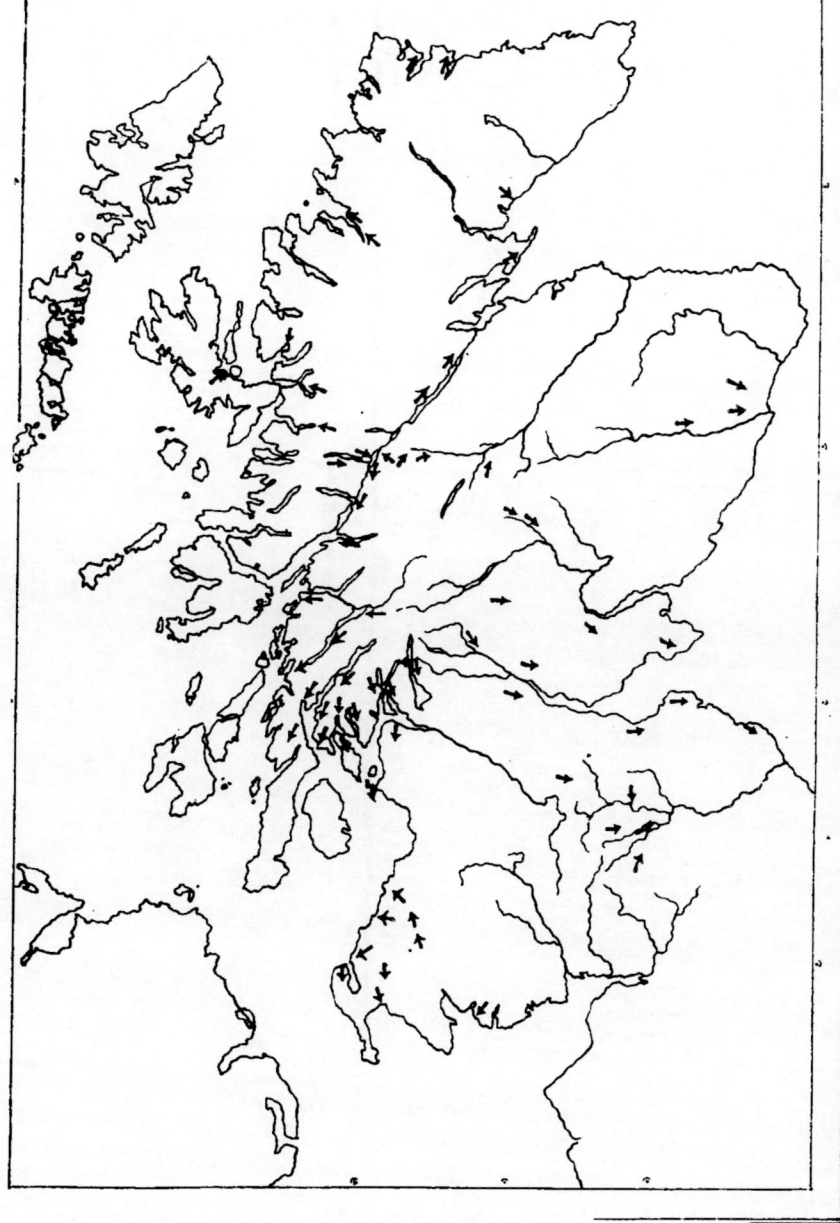

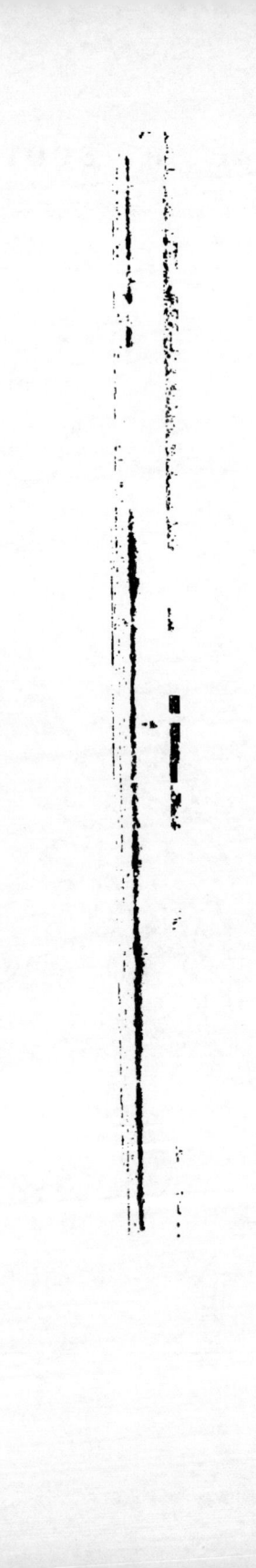

Plate I). Down all the western fiords they may be traced along the wavy undulating bosses of polished rock until they pass beneath the waters of the Atlantic. Along the Pentland Firth they may be seen in like manner descending from the high grounds of Sutherland northwards to the coast-line. On the eastern side of the island the same seaward trend of the ruts and striæ on the rock is traceable from Caithness to Berwick. Into the long valley of the Great Glen the striations come down from the high mountainous tracts on either side, and turn into the line of the valley, so as to run out into the Moray Firth on the north and into the Linnhe Loch on the south. In the glens that open upon the estuary of the Clyde the rocks are striated along the line of each valley, passing up inland into high grounds in the interior, and striking outwards beneath the sea. The very islands in that firth are striated in the same way. Bute, for instance, is a notable example, for the striæ, after coming down the glens of Cowal and passing beneath the Kyles, reappear on the Bute shores, actually mount the slopes of the island, so as to go right across it at a height of more than five hundred feet, and descend upon the firth on the south-west side. (Fig. 10.) Again, we can sometimes trace the striations out of one glen or sea-loch over a watershed into another of the numerous long and deep inlets that ascend from the basin of the Clyde. Thus from

Loch Lomond these strange almost indelible markings can be seen striking through the short cross valley at Tarbet and descending upon Loch Long at Arrochar. From the latter loch, again, they may be followed over the watershed which separates that fiord from the Gareloch, and thence down the latter valley into the Clyde.[1] In Loch Fyne, also, continuing in the line of the upper part of that valley, they are not deflected when this loch makes a bend

FIG. 10.—ICE-WORN ROCKS NEAR SUMMIT OF BARONE HILL, BUTE.

south of Ardrishaig, but actually ascend the hills above Tarbert, and cross heights of eight hundred feet into the Sound of Jura.[2] There is no great sea-loch, or glen on the west coast where these features may not be seen to a greater or less extent. As an

[1] First described by Mr. Maclaren. Edin. New Phil. Jour. vol. xl.
[2] Described by Mr. T. F. Jamieson. Quart. Journal, Geo. Soc. vol. xviii. p. 177.

example, at once easily visited and eminently characteristic, I may refer to the valley of Glendarual, which, descending from the mountains of Argyleshire, opens into the inlet of Loch Riden and thence into the Kyles of Bute. If the observer will take boat and row gently along the rocky shore, and among the numerous islets of the comparatively short estuary of Loch Riden, he will be amazed at the freshness in which the smoothed and striated rock-surfaces have been preserved. Wending his way between the islets and headlands, he will notice that on looking towards the head of the loch, his eye catches the rough blunt faces of the different crags and knolls; that, as he passes them, their sides, parallel in a general way with the sides of the loch, are well polished and striated, and that their upper ends, or those which face up the loch, are all worn down and smoothed off. He could not desire a more instructive lesson as to the nature of that smoothing process to which the surface of the country at large has been exposed. The striæ running parallel to the loch, the blunted and worn look of those parts of the crags and hummocks of rock that point up the valley, and the comparatively fresh and rough faces of those that look towards the open sea, indicate, as plainly as words can do, that the agent which smoothed and striated the sides of Loch Riden must have moved downwards along the length of the valley from the high grounds of the interior. Nor is this all. An ascent of the valley of Glendarual, above the head of the sea-loch, will show that the agent which produced the striations must have filled up the glen to the brim, and actually overflowed it. For the long parallel markings on the rocks can be followed as they slant over the west side of the valley and pass across the hills of Cowal into Loch Fyne. The ground between that loch and the Glendarual valley has been worn down into many hollows which, with the striations, trend away to the south-west.

Geologists are now generally agreed that these smoothed surfaces of rock, along with the striæ and grooves which cover them, have been produced by ice. Till lately the commonly received opinion in this country has been that they were caused, during a submergence of the land, by icebergs laden with sand and blocks of stone, whereby the rocks at the sea-bottom were scratched and worn down as the icy masses continued to be driven upon them. Within the last few years, however, this explanation has anew been called in question as inadequate to explain all the phenomena. The striæ, as we have seen, do not merely run along the top or side of a hill, they sometimes run up and over it. Moreover, they accommodate themselves to all the little inequalities of the surface over which they pass. This could hardly have been done by a rigid mass of ice moving horizontally like a berg or floe, with no determinate motion of its own, but driven by the ocean-currents and by winds: on the contrary, it points to an agent of such plasticity, as to be able to mould itself upon the irregularities of its rocky bed and to rise or fall as the nature of the ground required. And this agent, as shown by the divergence of the striations, must have moved outwards and downwards from the chief mountain masses, such as the Grampians and the chain of heights from Loch Eribol to the Sound of Mull. It must have filled up wide and deep valleys, pressing everywhere steadily and mightily upon the rocks, disregarding the minor features of the surface of the country, passing even over hills six or eight hundred feet high, as if they were but mole-hills, and continuing its operation over so vast a period as finally to leave the country smoothed and polished, or as it were, moulded into a flowing undulating contour.

To fulfil these conditions the only agent known in nature appears to be *land-ice* or *glaciers*. The polished and striated rocks find their exact counterparts along the course of every modern glacier. There is hardly a Highland glen, nay, strange as it may sound, there are not many hill-sides, even of the Lowlands, which do not remind one of the *roches moutonnées*, or ice-worn knolls of the Alps. The moulded outlines are the same, the striæ are the same, and the parallelism of the striations with the direction of the long valleys is alike the same, in Scotland and in Switzerland. In comparing the rock-markings of the two countries, we are driven to admit, that as in the one case we see these markings to be manifestly the work of moving glacier-ice which is still visible and still producing the same results; so, in the other instance, precisely similar effects can hardly but be due to the same cause, although the Scottish glaciers have long disappeared. It is more than twenty years since Agassiz, fresh from a study of the Swiss glaciers, announced this conclusion. But British geologists, after trying every other expedient, have only recently begun generally to adopt it. Their difficulty lay not in the admission of the existence of glaciers in Scotland, for admirable descriptions of glacier-moraines and striæ in Skye, Forfarshire, and Argyleshire, were published by Principal Forbes, Sir Charles Lyell, and Mr. Maclaren.[1] But if the universal striation were everywhere taken as evidence of the existence of land-ice, it was plain that Scotland must not merely have been the seat of local glaciers as Switzerland is at the present day, but must have been actually sealed in ice from mountain-top to sea-shore. This was a supposition

[1] See these and the other papers referred to in chronological order in Appendix II. to my Memoir on this subject. Mr. R. Chambers has long believed the whole of Scotland to have been under ice.

too violent for ready belief, and hence the attempt to account for the smoothing and striation of the country by iceberg action.

But the iceberg hypothesis must, I believe, be abandoned. Geologists are at length reluctantly, and against all their previous speculations, driven to confess that, after all, Scotland, and probably at least the north of England, must have been swathed in one vast wintry mantle of snow and ice. The present aspect of the Greenland continent, already referred to, probably gives us a close reproduction of the scenery of our own country when the cold of the glacial period was at its height. It is plain that such a thick mass of ice enveloping the whole land, and ceaselessly pressing down towards the sea to break off there into bergs, must have a constant grinding movement far grander in its geological results than that of any mere valley-glacier. The glacier wears down only the sides and bottom of the valley in which it flows; but the great Greenland ice-sheet, covering the length and breadth of the country, and allowing the underlying rocks to be seen only in occasional inland peaks, and in a narrow interrupted strip along the sea-coast, must effect an abrasion of the whole surface infinitely greater than any local glacier could do. We have no reason to suppose that the surface of Greenland differs from that of neighbouring areas of the northern hemisphere. The portions which every summer sets free from the snow, suffice to indicate that those parts which are never bared of their perennial wintry garb, partake of the same inequalities, here intersected by valleys, and there rising into ridges and hills, nay, even sweeping up into lofty peaks, which are sometimes seen protruding black and jagged above the snow. Yet these varieties in the contour of its bed do not prevent the steady seaward descent of the ice. Where the ice is much crevassed it may be travelling over rocky or steep ground, while in other places, where the ground probably slopes more gently, the surface of the ice is unbroken. But whether its floor be rough or smooth, steep or level, the ice is ever moving with a resistless march towards the sea. Its mass must thus be so immense that it treats as mere pebbles in its path, ridges and even hills of possibly very considerable elevation. It seems to pass as easily over them as a deep river sweeps over the stones that roughen its channel.

The Scandinavian peninsula affords an interesting connecting link between the existing condition of Greenland and the state of Scotland during the glacial period. The immense snow fields of Norway are but the relics of the sheet of ice and snow which once covered that country and turned it into another Greenland. This sheet has shrunk up into the high grounds, from which it protrudes into the valleys in the form of glaciers, but it has not done so without leaving its stamp on the surface of the country The rocks are there also glaciated, or worn into those flowing outlines which have been described as characteristic of the rocks of Scotland. Down the valleys and fiords may be traced the polished surfaces, with their ruts and grooves still as fresh as if they had only lately been made. From the sea-margin, where these markings slip under the waves, and where no ice is now to be seen, we can follow them upward step by step, till they pass into those which the great glaciers are now graving on the rocks of the interior. The inference cannot be withheld that at one time these glaciers, instead of melting away where they now do, stretched far down the valleys and went out to sea, and that as the striæ can be found high on the hill sides, the valleys must have been filled to the brim with ice. In short, Scandinavia at that time was in the same state in which Greenland still remains. The climate has grown milder, indeed, but the great inland snow-fields and glaciers yet continue as silent witnesses of the severity of the ancient climate.

Passing southwards, we see the perpetual snow gradually disappear, and when we reach the British Isles, it is gone, although in sheltered crannies of the Grampians gleaming patches may be seen late in autumn which have outlasted the summer and will lie till covered by the drifts of the next December. And yet, though snow-field and glacier have fled, they have left their mark as clearly and widely on the rocks of Scotland as on those of Norway. A calm investigation of the rock-markings seems to me to render this conclusion certain. No iceberg action could have moulded the contour of this country to its present form, any more than icebergs could have worn down the hills and valleys of Scandinavia. The more we pursue this supposition the more utterly inadequate does it appear. That icebergs may perform a good deal of abrasion as they grate along the sea-bottom, and that much may have been done in this way in the British area, more especially over the Lowlands, during some parts of the glacial period, no one will probably deny. But the divergence both of the rock-mouldings and of the striation from the main mountain-ranges shows, I think, conclusively that the force by which these changes on the surface were produced, did not come from a distance, but had its seat in the country itself. It is quite possible, indeed, that when, during the glacial period, the land had sunk two thousand feet or more below its present level, fleets of arctic bergs may have drifted over the submerged hills and scored them with ruts and grooves running from north to south. By this means it may be just to account for such high-lying groovings and striæ (first noticed by Mr. Chambers), as cross what must have been the direction of the flow of the ancient ice-sheet. Floating ice, also, derived from the Highland glaciers, may have drifted out to sea and have rasped the sides and summits of the less elevated Highland hills, as well as the heights in the low country. But this explanation leaves by far the larger part of the glaciation of the country unaccounted for. The striation of Scotland, down even to its minutest features, is similar to what is now effected by the moving of a body of land-ice, such as a glacier. In Switzerland the evidence of a former greater extent of the glaciers rests upon the height and distance to which the striation, moulded surfaces, and travelled blocks can be traced beyond the present limits of the ice. And this evidence is of such a kind as amounts to a demonstration. In Norway the surface of the country between the snow-fields and the sea presents the same rock-mouldings and groovings as are even to this day produced by the movement of its glaciers. In this case, also, these markings are demonstrably the result of moving land-ice. The conclusion therefore can hardly be resisted that the very same markings over the surface of Scotland must have been made by the same agency, and thus that this country must have been covered with ice and snow, which passing from the high grounds outwards and seawards for ages, wore down the rocks and gave the land that undulating and truly ice-worn aspect which it wears to this day. In short, our country in old times must have been very much what Greenland is now—a land of perpetual snow and ice, pushing from its numerous fiords and glens huge glaciers to the sea.[1]

Possible thickness of the ice. It is still possible approximately to estimate the thickness which this cake of ice reached in some of the Scottish valleys. Thus, in Loch Fyne, both sides of the valley are smoothed and striated; nay, the whole of the land between that Loch and the western sea on the one side and Loch Long and the Firth of Clyde on the other, bears the same evidence of vast abrasion. Striæ have been observed by the Duke of Argyle on the hill-tops at a height of 1,800 feet above Loch

[1] The cause of the cold of the glacial period has been a fruitful source of speculation. That the phenomena of striated rocks, moraine mounds, and transported stones are so wide-spread over the globe has long seemed to me a strong presumption against the attempt to explain the cold by mere changes in the distribution of land and sea. Agassiz, who has been so far ahead of British geologists in this department of the science, saw so long ago as 1842 what some of our more eminent geologists do not seem to see yet, for we find him asking, "who would reject the idea that the cause of this cold has been general, and attribute to local causes effects so diffused over the surface of the globe?" (*Edin. New Phil. Journ.* xxxiii. 240.). The true explanation of this confessedly difficult subject is, I believe, that given recently by Mr. Croll, who accounts for the recurrence of cold and warm periods in the geological history of the globe by changes in the eccentricity of the earth's orbit. His paper (in the *Philosophical Magazine* for August, 1864) is one of the most important contributions which have been made to geology for many years. Among its fruitful results will probably be the key to the value of geological time.

Fyne, running parallel with the valley like those at a lower level. The greatest depth of the loch is 624 feet, and as the whole sides and bottom were probably striated in the same way and by the same agent, there is reason to believe that the ice, even if the higher markings were produced some time before those at the bottom,[1] was probably not less, but possibly much more than 2,000 feet. Mr. Maclaren has traced the striæ up to heights of more than 2,000 feet in the south-west Highlands. Mr. Jamieson also estimates that the ice in Glen Spean must have been two miles broad at the surface, and at least 1,300 feet deep.

And what has been the result of the extension of this thick wide icy mantle over the country? We

---

[1] A very necessary qualification, for it should not be forgotten that the ice must have been ever grinding down its bed and deepening the valleys. The markings that run high along a hill-side might thus have been made when the ice was at a considerably higher level than the present bottom of the valley below, while those along the bottom may have been engraved at a far later time when the ice had ground its way down through the solid rock so as greatly to lower the level of the valley. Hence it is not quite safe to say absolutely that because we have striæ 2,000 feet above the floor of an old glacier-glen, the ice must have been more than 2,000 feet thick, for this estimate makes no allowance for the lapse of possibly a long interval, and an extensive erosion of the valley between the striation of the hill-tops, and that of the rocks at the bottom. Still, to smooth and furrow the ridges at a height of 1,800 feet above the sea, the ice must have covered them to a considerable thickness. Hence, perhaps, the number of feet to be added for the ancient height of the ice, may more than counterbalance the number to be subtracted for the subsequent lowering of the level of the valley. Such estimates can only be approximate in the meantime until some definite information is obtained regarding the actual extent to which the valleys were deepened by the ice.

have seen that the effect of ordinary atmospheric waste is to wear down the surface of the land, but at the same time to roughen and scarp it, loosening the joints and fractures of the rocks, cutting out ravines and valleys, carving high cliffs and precipices, and mottling the mountain-sides with many a rugged scar and crag. It was on such a surface that the ice of the glacial period began to settle down. But ice in motion acts differently from rains, springs, and frosts. Its tendency is to wear down the surface over which it passes, to rub away the roughnesses, to round off the prominent edges, to give a wavy flowing outline to the hills, and to smooth, deepen, and widen the valleys. And such have been its effects upon the surface of the Scottish Highlands. The glens carved out by that long process of subaërial waste already described, have been afterwards filled with ice. They served to drain off the ice as they had previously served to drain off the water. They have thus had their sides and bottom worn down and their rocks smoothed, polished, and striated. And even where the rocks are fast mouldering away under the renewed attacks of the subaërial agencies of waste, the footprints of the ice are seldom so utterly obliterated as not here and there to reveal their traces. The angularities of the pre-glacial land have indeed been effaced, and their place has been taken by the undulating contour produced by the great ice-sheet.

But the rains and rivers, springs and frosts are busily engaged in breaking up that smoothed surface. The ancient roughnesses are coming back again, and thus arises that mingling of rugged crag and scar with a well defined flowing outline, which forms so marked a feature in the scenery of the hilly and mountainous tracts of Scotland.

When we try to sketch to ourselves in imagination pictures of the changes which geology reveals, no matter how vaguely or hesitatingly the lines may be drawn, we are arrested by a feeling of inexpressible wonder at the contrast which is thus often brought before us between the present and past conditions of the country. To sit, for instance, on one of the headlands of the Firth of Clyde, and watch the ships as they come and go from all corners of the earth; to trace village after village, and town after town, dotting the coast line far as the eye can reach; to see the white steam of the distant railway rising like a summer cloud from among orchards, and cornfields, and fairy-like woodlands; to mark, far away, the darker smoke of the coal-pit and the iron-work hanging over the haunts of a busy human population; in short, to note all over the landscape, on land and sea, the traces of that human power which is everywhere changing the face of nature;—and then to picture an earlier time, when these waters had never felt the stroke of oar or paddle, when these hill-sides had never echoed the sound of human voice, but when over hill and valley, over river and sea, there had fallen a silence as of the grave, when one wide pall of snow and ice stretched across the landscape; to restore, in imagination, the dull sullen glacier threading yonder deep Highland glen, which to-day is purple with heather and blithe with the whirring of grouse and wood-cock; to seal up the firth once more in ice as the winter frosts used to set over it, and to cover it with bergs and ice-rafts that marked the short-lived Arctic summer; to bring back again the plants and animals of that early time; and thus, from the green and sunny valley of the Clyde, with all its human associations, to pass at once and by a natural transition to the sterility and solitude of another Greenland, is an employment as delightful as man can well enjoy. The contrast, though striking, is only one of many which the same district, or indeed any part of the country, presents to a geological eye. And it is the opening up of these contrasts, based as they always must be upon a careful and often a laborious collection of facts, which entitles geology to be ranked at once among the most logical and yet amid the most imaginative pursuits in all the wide circle of science.

Origin of Highand Lakes.

To the changes that went on during the glacial period must be attributed another characteristic element in the landscapes of both of the Highland and Lowland tracts of Scotland—the existence of lakes. Did it ever occur to the reader to account for the formation of a lake-basin? Perhaps he never thought of the subject, or if it entered his mind at all, he possibly settled the question by recalling the old Scottish proverb, "there never was a heich, but there was a howe." He will find, however, that the answer is often not to be reached without a good deal of labour. Lakes, at least those which mottle the surface of Scotland, may be grouped into three classes: 1st. Those which lie in original hollows of the superficial drifts. 2d. Those which have been formed by the deposit of a bar of drift across a valley or depression. 3d. Those which lie in a basin-shaped cavity of solid rock.

The whole of these appear to be due to the glacial agencies of the last great geological period. Those

FIG. 11.—LAKES LYING IN HOLLOWS OF DRIFT.

of the first class are usually of small size. In the Highlands and in the uplands of the south they are chiefly to be found filling the little cup-like hollows of old glacier moraines. In the wide lowland valley they lie in depressions formed by the unequal deposition of the boulder-clay or the upper sandy drift.

The second group is commonly exemplified in the mountainous and hilly regions, where the moraine of a glacier has been thrown down across a narrow glen or opener valley, and has ponded back the water of the streams. At the outflow of the lake thus formed

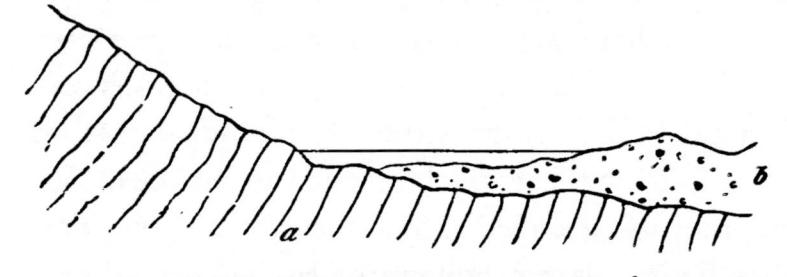

FIG. 12.—LAKE DAMMED UP BY RIDGE OF DRIFT. [See also Fig. 20.]
a. Underlying rocks; *b.* Ridge of drift.

the barrier of loose drift is cut down by the escaping stream, which year by year deepens its bed, and will gradually lower the lake until at last it drains it altogether.

In neither of these cases is there any obstacle to our seeing how there should be lakes. But in the third kind of lake-basin the difficulty of explanation is greater than at first sight appears. We have, then, to account for the existence of a cup-like or basin-shaped hollow in solid rock, sometimes along the line of a valley, sometimes on a plateau, sometimes on a hill-top or on a watershed. As mentioned in a previous chapter, Professor A. C. Ramsay was the first to point out what I believe will be ultimately accepted

Highland Lakes. 173

as the true origin of these hollows. He has shown, that as they are not helped by deposits of drift, but lie in naked rock, so they are not due to rents, or corrugations, or depressions of the earth's crust, but have been actually scooped out of the rocks. They are, as I have already said, hollows of erosion, just as

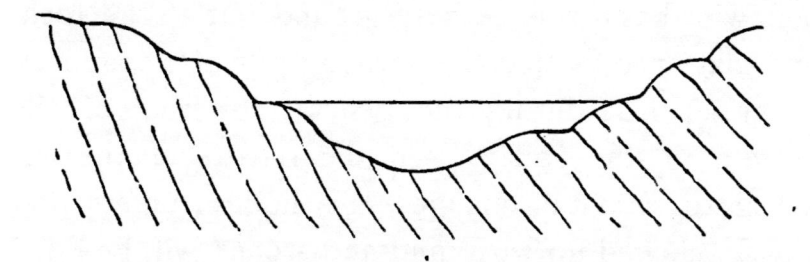

FIG. 13.—LAKE BASIN IN HOLLOW OF SOLID ROCK.

much as the river valleys. What agency could have produced them? Not running water certainly, nor the waves of the sea, nor rains, springs, and frosts. Mr. Ramsay has assigned the only cause which we at present know capable of eroding such hollows, when he attributes their origin to the excavating power of land-ice.[1] A river of ice is not bound by the same restraints as those which determine the action of a river of water. When a glacier is, as it were, choked by the narrowing of its valley, the ice actually rises.

[1] It is perhaps conceivable that the great fields of drift-ice with enclosed bergs which annually move southward from the Arctic seas may, by grating along the sea-floor, produce occasional hollows comparable on a small scale to those which are regarded as the result of the long-continued action of land-ice.

In such places there is necessarily an enormous amount of pressure, the ice is broken into yawning crevasses, and the solid rocks suffer a proportionate abrasion. The increased thickness of the mass of ice at these points must augment the vertical pressure, and give rise to a greater scooping of the bed of the glacier. If this state of things last, it is plain that a hollow or basin will be here ground out of the rock, and that once formed, there will always be a tendency to preserve it during the general lowering of the bottom of the valley. On the retreat of the ice, owing to climatal changes, this hollow, unless previously choked up with mud and stones, will be filled with water, and form a lake. It will be a true rock-basin, with ice-worn surfaces around its lip, and over its sides and bottom.

And such is really the appearance of many a lake and tarn in the Highlands of Scotland. One of the largest and noblest of the whole—Loch Awe—may be taken as an illustrative example. The present outflow of the lake through the deep narrow gorge of the Pass of Brander is comparatively speaking recent. It has been opened across the lofty ridge that stretches from King's House through Ben Cruachan to the Sound of Jura. I regard it as another example of a watershed cut through by streams which flow in opposite directions, aided doubtless both by the sea and by the stream of ice that came down from the opposite mountains, and pressed through every available outlet to the ocean. But no one can ascend from the Sound of Jura to Kilmartin, and thence up the terraced valley to Loch Awe, without being convinced that this must have been the old outlet of the great valley of that loch. The drainage from the cluster of long deep glens at the head of the lake came down into the main valley and went out to sea in the Sound of Jura. The excavation of the valley into a long lake-basin, and the cutting of the watershed through the Pass of Brander are later events, both probably dating from the glacial period, while the origin of the main valley of Loch Awe carries us back infinitely far into the past. It is owing to this erosion of the valley and to the cutting through of the Brander Pass that the water now flows into Loch Etive instead of the Sound of Jura or Loch Gilp. One cannot but wonder when, on ascending the valley from Kilmartin, he at last finds himself on an ice-worn barrier of schist, and sees stretched out for miles before him the wooded shores of Loch Awe. The lake is dammed back by hard rock, yet the smoothed and polished surface of the barrier and the parallelism of its striations with the length of the valley, show that the mass of ice which once filled up the present basin of the lake passed on down the continuation of the valley towards Kilmartin. And all along the sides of the loch, and on its rocky islets, the same traces may be seen of the steady southward march of the ice. The rocks are worn into smooth mammillated outlines, and covered with ruts and grooves that trend with the length of the valley. It is, in short, a rock-basin of which all that can be seen is ice-worn; and if further proof of the old glaciers were needed, it would be found in the heaps of moraine rubbish piled along the side of the valley.

All the Scottish rock-enclosed lakes lie, I believe, in ice-worn basins. In certain districts, particularly over the area occupied by the Laurentian gneiss, such as the island of Lewis, the surface of the ground has been, as it were, sown broad-cast with tarns, lochans and lakes—lying in dimples and depressions worn out of the rock. But throughout the Highlands one cannot fail to note that the larger sheets of water all lie along the glens in the paths of old glaciers. Like the glens, they are arranged with reference to the drainage of the country; and just as it is inconceivable that the country could have been rent originally into the pattern of its drainage system, so it is impossible to admit that the lakes can be due to a series of rents and subsidences which in each case were neatly adjusted to the size and direction of the valley. These lake-basins are truly hollows of erosion. The conditions under which they were scooped out may have greatly varied, depending now upon the form of the ground, now upon the nature of the rocks, and sometimes perhaps upon causes which are not at present evident. Occasionally the lakes must coincide with a line of fault, just as the glens do. But they are not on that account to be set down indiscriminately as the work of subterranean commotions. The most striking example of this coincidence is shown in the line of the Great Glen. That singular depression, cutting Scotland into two, runs probably along a line of fracture throughout its entire length.[1] It will not be denied that the trend of the glen has been defined by that of the fracture underneath. It may be further insisted, however, that the sides of the glen are themselves the sides of the open fissure, worn down indeed, by subsequent

[1] Its very straightness is enough to suggest that it owes its direction to this cause. I ascertained last autumn that the effects of the fracture, or of one continued in the same line, can be seen along the western side of the Moray Firth where the Jurassic beds of Eathie and Shandwick are thrown down against the Old Red Sandstone. Hence the downthrow at this end of the line is to the east side. It seems to me that this line has been from a very early geological period up, indeed, to the present day, a line of weakness in the crust of the earth. The prolongation of the tongue of Old Red Sandstone up the valley of Loch Ness appears to show that the valley is older than that formation: the dislocation of the Eathie and Shandwick shales proves disturbance even after the Lias; and the agitation of the waters of Loch Ness, during great earthquakes in modern times, shows that even yet underground movements tend to reveal themselves along the same old line. Hence it may be reasonably conjectured that the fracture along the line of the great glen has been repeatedly modified during the subterranean changes of successive geological periods.

denudation, yet still gaping asunder as when they were first parted. But an examination of the valley will convince the observer that the amount of rock which can be shown to have been worn away would have sufficed to fill up the supposed cavities. Let him, for instance, stand on the ice-worn barrier of rock between Loch Ness and Loch Oich. He will see there that even on the supposition of an open fissure the deep concavity of the glen at this point must be due to denudation, for as the rocks can be traced across the bottom of the valley there is no room for a wide open fissure. The very arrangement of the rocks is enough to prove, that the hollow has been worn out by the erosive agencies of nature; the upturned strata, vertical or highly inclined, present their truncated ends to the sky, and can be followed bed after bed across the glen till they rise high into the hills on either side. These bared ends, of course, were once prolonged upward: no fissure or fault could have exposed them; the lost parts can only have been removed by some agent which acted, not vertically like a fracture, but in a general sense horizontally, such as rains, rivers, frosts, ice, and the sea. In short, the glen at Fort Augustus must be due mainly to denudation; the direction of the erosion being determined originally by the feature which the long line of crack made at the surface. And if this be the case at one part of the glen where both its bottom and sides can be examined, it is plain that the same structure can hardly fail to characterise the rest of the valley where it is filled by the lakes. The same evidence of extensive denudation, indeed, can be followed along the sides of the lakes. There cannot be any doubt that since the dislocation began the Great Glen has suffered to a vast extent from the inroads of time. The material worn off its sides would doubtless first find its way to the bottom, and if the lakes had existed they would have been far more than filled up. The lakes are therefore later than the formation of this long valley. It seems to me that this conclusion must be conceded even by those who most strenuously oppose the erosive power of ice.

If then the deep cavities of the Great Glen [1] are of later date than the scooping out of the valley itself, to what source must their origin be traced? There seem to be only two possible answers to this problem; the lochs must either be due to special fractures or subsidences after the formation of the valley in which they lie, or they have been dug out by ice. The question is thus narrowed to the same issue as that which every rock-enclosed lake proposes to the geologist. It may perhaps be impossible to prove that a valley, which had its course defined at first by

[1] The deepest sounding in Loch Ness gives a depth of 129 fathoms opposite the Falls of Foyers; in Loch Oich, 23 fathoms; in Loch Lochy, a little below Letterfinlay, 76 fathoms.

a rent of the earth's crust, and which since then, up even to recent times, has been subjected to subterranean tremors, has not been rent open into deep lake-basins at a late geological period. But to affirm that it has been so sundered is a mere assertion without proof. So far as the evidence of the ground goes, it seems to me to bear strongly in favour of Professor Ramsay's doctrine that the hollows are the work of glacier-ice. The Great Glen receives the drainage of a wide mountainous region on either side, and in old times a larger amount of ice probably flowed into it than into any other valley in Scotland. It received from the west the large glaciers of Loch Eil, Loch Arkaig, Glen Morriston and Glen Urquhart, from the east those of the Glens of Lochaber, and those which come down from the north-western flanks of the Monadhliah mountains. Its sides show everywhere the flowing rounded outlines that mark the seaward march of the ice, and its rocky bottom, where visible, bears the same impress. That it has been ice-worn is thus rendered plain. Were there no other examples of such ice-worn cavities in Scotland it might be too bold to affirm these lake-basins of the Great Glen to have been scooped out by ice. But when the same features can be seen in hundreds of other instances where no fracture or subsidence can be shown to exist, this view of their origin seems in the highest degree probable.

I have been thus particular in referring to the chain of lochs along the Great Glen, because they, more than any other Scottish lakes, appear to bear out the hypothesis of subterranean fracture and disturbance. And if it can be shown that even they do not prove this hypothesis, but, on the contrary, point to the grinding power of ice, some advance is made in the explanation of the history of this characteristic feature in the scenery of Scotland. At the risk of tediousness let me sum up in one paragraph the argument in favour of the theory of glacial erosion.

All the Scottish lakes, without exception, lie on a surface that has been enormously denuded. Those that are held in rock-basins exhibit along their margin, on their islets, and on their bottom, as far as it can be examined, the same polished and striated rock-surfaces which are found universally throughout the kingdom.[1] These basins appear to be as thoroughly moulded and striated from end to end as any part of the country. It is admitted that this polishing and grooving of rock-surfaces has been produced by the grinding action of sheets of solid ice. It follows that the polishing and grooving of the Scottish lake-basins has been the work of ice. But this process could not have been accomplished unless the ice had actually gone down into the basins

[1] The above summary of the argument was used by me in the *Reader* for 4th June, 1864.

and grated along their bottom. That it did so seems proved beyond dispute by the grooves and striæ which can be traced slipping under the waters of a lake, rising and sinking again over the surfaces of the islets and submerged bosses of rock, and finally re-emerging with the same steady line of bearing from under the water at the farther end. But, if ice could do this, if it could score and furrow the rock of a lake-bottom from end to end, it must, in so doing, have helped to deepen the basin: and if ice can deepen a basin, it can make one entirely. There must indeed be a limit to this erosion when, for instance, the pressure behind is no longer able to drive the ice out of the hollow, or beyond its lower end. But that up to such a point ice is a powerful excavating agent I cannot doubt. Weighing the matter well, it seems to me that the origin of the rock-basin lakes of Scotland is inseparably interwoven with the general glaciation of the country, and that geologists must either reject the theory of such a general glaciation, with all its cumulative proofs, or admit that these lake-basins lie in hollows that have been mainly scooped out by ice.[1]

When this view of the origin of these rock-basins

---

[1] There may indeed be difficulties in the way of explaining how ice can mount a slope in the manner in which it is held to have done. But with the long striations and deep furrows before me and the moulded ridges all running up hill, I cannot but believe that the ice actually did ascend, and in so doing, wore down the rocks.

is taken, a new light is cast upon the history of the western sea-lochs. These deep inlets, as I have tried to show, are probably submerged land-valleys. Some of them, Loch Fyne, for example, are in part deeper than the sea immediately outside them, as Loch Ness is deeper than the German Ocean. These abysses, if the land were sufficiently upheaved, would be marked by lakes lying along the valleys. Were Loch Fyne raised in this way above the sea-level, it would be like any wide inland glen or strath of the present day—a deep, broad valley with tributaries entering from the hills on either side, and with a stream in the centre flowing through one or more lakes. Nay, more, its sides and bottom, so far at least as they can be seen at the present time, would be found smoothed and striated like all that now rises above low-water mark. These markings are unquestionably the work of ice, and, if so, there is, I think, a high probability that the deep basins in the western fiords have been scooped out by the same powerful agency. But the land in that case may have had a considerably greater altitude than it has now, to enable the thick ice-sheet to fill and wear down the wide valleys that are to-day washed by the tides of the Atlantic. And this conclusion in favour of a much higher elevation of Scotland during a part of the glacial period, tends to confirm the inference already drawn from the fact that it is the prolongation of the glen into the sea which gives rise to the fiord.

Thus, by the evidence of the rounded and worn outlines of the hill-sides and glens, the smoothed and striated surface of the rocks, and the frequent occurrence of rock-enclosed tarns and lakes, we learn that the extent to which this country suffered from the abrading action of its ancient land-ice must have been enormous.

As the amount of rock worn away was so vast, the quantity of detritus produced must have been proportionately great. The rivers which issue from Alpine glaciers are thickly charged with the fine mud made by the friction of the sand and stones borne along under the ice: and in the Arctic regions the sea is sometimes discoloured for miles from shore by the mud ground down from the surface of the land. In Scotland, too, since we have such abundant evidence of abrasion, there can be no doubt that there was a corresponding production of detritus. What, then, has become of it? If the rock-mouldings, and striation and the endless rock-basins, are enough to prove the passage of a massive sheet of ice from mountain to sea, it will add not a little to the impressiveness of that testimony if we can still point to the detrital matter which the ice left behind it. Happily this is not far to seek.

Débris made by the great Ice-sheet.

Boulder Clays. — Away from the mountains in such flat tracts as Caithness, the shores of the Moray Firth, with those of its tributary inlets, and the low land bordering the German Ocean, the solid rocks are to a large extent covered with a stiff clay full of stones varying in size up to boulders that measure, occasionally, a yard or more in diameter. To this deposit the name of *boulder-clay* or *till* has been given. It attains its chief development in the great Lowland valley. Hence, perhaps, its detailed description had better be deferred until the surface of that valley comes before us for examination. Its origin has long been involved in mystery. The older geologists called it diluvial, and regarded it as a proof of a violent flood, or a series of floods, which, sweeping across the country, produced the striation on the rocks by driving over them the stones, sand, and clay which now form the Till. The introduction of land-ice into the list of agencies concerned in changing the surface of the country has now, however, given us the clue to the history of this remarkable superficial deposit. As will be afterwards pointed out, its internal structure, and its striated stones, show it to be the result of the abrasion carried on by the ice-sheet as it moved over the land.

The scenery of the districts where boulder-clay prevails, though tame and monotonous, has certain characteristic features of its own. The surface is usually smooth and undulating, save where some knob or hill of the rock below rises rough and broken above the soil. It is along a sea-cliff, or better still, along the sides of a river, that the more prominent features of the deposit can best be seen. It there forms a line of steep green bank, projecting at frequent intervals into massive turf-covered buttresses, and receding in the spaces between into grassy hollows furrowed here and there by runnels from the level or sloping ground above. It is by the alternate influences of rains and frosts that these strange rampart-like slopes have been made, and the process may be seen still in progress, for every winter loosens here and there a mass of the clay which, slipping down the bank, leaves a semicircular scar of raw clay above, as if a huge spadeful had been dug out of the slope. The fallen portion, with its covering of turf, continues to slide as the water trickles below it, but at last, if it does not reach the stream below it becomes so far shielded by the coating of grass and weeds which creeps over it. It then remains one of many verdurous mounds and hummocks, mottling the sides and base of the declivity like the ruins of a set of earthworks older than the steep *glacis* that rises behind them. A sloping bank of boulder-clay under the wasting hand of time thus assumes a strangely uneven outline, and when the action which gave rise to these features ceases for a time, and the irregular surface puts on everywhere a covering of turf which gets trodden into narrow undulating terraces by the sheep, the result is often a confused series of mounds and ridges, so artificial in their general appearance as to give rise to legends of giant's graves, and fairy knowes, or to traditions of ancient camps, and forts, and tumuli. The influence of the boulder-clay as an element in the landscape may be well seen along the sides of the Cromarty Firth. It forms there a smooth slope from the side of the bare high grounds down to the inner edge of the platform of the last raised-beach, where it runs as a sinuous grassy bank worn here and there into some of its characteristic and fantastic forms.[1] From the Cromarty Firth eastward to the headlands of Aberdeenshire, the strip of low land between the hills and the sea owes much of its fertile character to its covering of boulder-clay and gravelly drift. A prevailing evenness of surface marks where the rocks are concealed by those deposits, which further reveal themselves on the sea-margin in lines of the same green rampart-like slopes, and along the river-courses

---

[1] See Hugh Miller's description of this scenery in his *Schools and Schoolmasters.*

in steep banks or *scaurs*. From Inverness, northward, masses of boulder-clay are found here and there along the coast-line. It forms high cliffs at Rosemarkie. At Tain, too, it rises in a line of bold bluffs, that mark the limits of an ancient shore. At the Ord of Caithness it is seen capping the headlands of granitic rock, but northward, where the mountainous tracts of Sutherland sink into the plains of Caithness, the boulder-clay spreads far and wide over the surface. At Wick it is largely developed, and it has there long been known to contain marine shells.

Higher Elevation and subsequent Submergence of the Highlands.

The boulder-clay is a deposit of the lower grounds rather than of the hilly and mountainous regions. As it now remains it is characteristically a lowland formation, and this arises partly from the circumstance that it must always have been more thickly developed in the low grounds than among the mountains, and partly because such parts of it as were laid down among the high grounds have been in large measure cleared away. When the boulder-clay began to be formed, the land probably rose considerably higher out of the sea than it does now, and there can be little doubt that, as the formation of that deposit went on, this old ice-covered land

Elevation and Submergence of the Highlands. 189

was slowly settling down into the ocean. It has already been pointed out that the sea-lochs of the western coast give evidence in favour of a greater elevation of the country at the time when these valleys were worn down by the ice. How much higher it may have been than it is at present we do not know. Perhaps the land was not only higher as a whole, but special parts of it—such for example as the western half of the Highlands—may have had a greater altitude relatively to the surrounding country than they have now. Unequal oscillations of level would help to remove several geological difficulties of a local kind which occur when we try to conceive of the movement of a large body of ice over such a surface as the country wears at present.[1]

If the evidence for the extreme height of the land during the glacial period is inconclusive, that for the extreme depression is equally so. In Wales the submergence was not less than 2,300 feet beneath the sea level, and in the southern half of Scotland it may have been as much as 2,000 feet. The difficulty of

---

[1] The notable example of the depression of Wales to a depth of more than 2,000 feet while the south-east of England was not submerged, should not be forgotten. We are too apt, unconsciously, to regard as permanent the present relative levels of different parts of the country to each other, and to think of the land rising or sinking as a whole; though we may be ready enough to admit when required, that there may have been unequal and partial elevations in recent geological times, one region remaining at rest, or actually sinking, while other parts were undergoing upheaval.

fixing the former limits of the sea upon the Highland mountains arises from the fact that when the land began to rise again, it was still cased in ice, which, pushing its way out to sea, would of course destroy the beds of marine sand, gravel or clay, from which, had they been spared, the old sea-margin might have been traced. So that although there is no reason to doubt that the area of the Highlands participated in greater or less degree in the general sinking of the British Isles, conclusive proof of the inference is not easily found.[1] Perhaps the most satisfactory evidence is to be drawn from the occurrence of large erratic blocks of granite, gneiss and schist, lying high on the sides and on the summits of the Sidlaw and Ochil Hills, and even on the rising grounds to the south of the Forth. These boulders were probably carried by icebergs which, breaking off from the ends of glaciers that came down to the sea-level from the heart of the Highlands, were drifted southwards, and dropt their mud and stones over the submerged lowlands of Scotland. They will come before us again when we trace the history of the great midland valley.

[1] Neither the occurrence of perched boulders, nor of mounds and terraces of drift, is of itself sufficient to prove a former sea-level, unless there be other corroborating evidence. For boulders among mountains might be carried and left by land-ice, and drift arranged by fresh water could not, I suppose, in the absence of fossils, be certainly distinguished from a marine deposit.

Second Glacier Period in the Highlands.

Such parts of the country as remained above water when the submergence was greatest, continued in all likelihood to be covered with snow and ice. And when the land began once more to rise out of the deep, the same thick wintry mantle, probably for a time, still wrapped it round. As the elevation went on, however, and the climate grew less severe, the massiveness of the ice-sheet became impaired. Instead of covering the whole country as it had done before, it now shrank up into the uplands, whence long glaciers threaded their way down the glens. But though its dimensions after the submergence were lessened, it yet remained in such force that from some of the higher groups of mountains, glaciers still descended to the sea until the land had risen to within at least forty feet of its present level. These ice-rivers in place of creeping onwards over an arctic surface, where the rocks were everywhere buried beneath snow and ice, were now confined between the sides of valleys and glens whose bare rocky slopes often rose high above the glacier that crept along their bottom. Mud, stones, and huge blocks of rock, disengaged by springs and frosts from the heights, rolled down upon the ice, which carried them slowly down the valley, and, where it finally melted, heaped them together into irregular mounds or *moraines*. The rocks over which the ice moved were, of course, subjected to a renewed process of abrasion, and thus a new series of *roches moutonnées* was produced.

There are few minor features in the scenery of the Highlands better fitted to impress the fancy than the memorials of these last remnants of the ancient ice. Huge mounds, cumbered with blocks of every size, some of them larger than many a Highland cottage, extend across a glen as if to bar all attempts to penetrate into its recesses. As these ramparts sometimes stretch without break from one side of the glen to the other, and rise to a considerable height, the traveller who picks his way up the valley among scattered blocks of rock and ice-worn knolls is puzzled to conjecture what may lie behind the rugged boulder-covered barrier of detritus that rises so formidably in his front. He mounts its outer slope, and on reaching the summit, sees below him, perhaps, a loch or tarn. The barrier is, in short, a glacier moraine which has been thrown across the glen; and though the ice has long since retreated, this bank of ice-borne rubbish has been left behind, serving to dam back the drainage of the upper part of the glen. A lake has thus been formed whose surplus waters are now laboriously cutting for themselves a pathway through the moraine. The time will come when the stream will have dug its channel as deep as the bottom of the lake, which will then be emptied, leaving a broad, flat meadow to mark where it stood. In many a Highland glen it is easy to trace the successive backward steps of the ice as it continued to shrink up into the higher recesses of the mountains. Each moraine shows, of course, a point at which the lower end of the glacier continued for a while stationary, melting there and throwing down its accumulated piles of rubbish. These moraines may be followed up the valley, mound within mound, each of which represents a pause in the retreat of the glacier, until at last we gain the upper end, where the stream of ice finally shrank up into the snow-fields, and where these, as the climate grew warmer, at last melted away.

I have said that the ice after the submergence was not so extensive as it had been before. But it still existed in such mass as to choke up many deep and long valleys, and powerfully to grind down the rocks over which it moved. The evidence of its magnitude and effects cannot be better seen in any part of the Highlands than among the glens to the north and east of Ben Nevis. Let the summer traveller who is in search of the picturesque, and who cares to follow the traces of ancient geological changes, ascend from the shores of Locheil up the valley of the Spean into the wilds of Lochaber. Quitting the shore, with its fringing terrace that marks a former limit of the sea, when the land was about forty feet lower than it is now, he enters Glen Spean among mounds of *débris*. These heaps probably mark where a glacier threw down its load of earth and stones either in the sea or in a former freshwater lake. As he gradually advances up the valley he finds this old glacial detritus cut here and there into terraces, level as a meadow, and contrasting strangely with the ruggedness of the surrounding ground. He is reminded perhaps of some of the broad alluvial haughs so common along the banks of the larger rivers. But these terraces, in the lower part of Glen Spean, rise high above the stream, and must be due either to the action of the sea, or, more probably, to that of a lake once filling the glen. As he approaches the foot of Glen Roy, however, his eye falls on a still more wonderful terrace. Far away up the valley, if the day be a favourable one, he can trace a line, carved along the steep hill-side to the south of the river, and running with seemingly mathematical precision in a horizontal course till it is lost in the distance. Before turning up Glen Roy he would do well to continue the ascent of Glen Spean, that he may see the marvellous proofs of the magnitude of the Scottish glaciers even after the submergence of the country. A short way beyond Bridge of Roy Inn the road, after running so long over the re-assorted glacial rubbish, begins to rise over hummocks of hard schist. These, instead of presenting the jagged, rough surface which they would have had if they had suffered only from rains and frosts, are rounded, smoothed, and polished, and they show in perfection the long, close, parallel striæ and scorings already described. They are in short examples of *roches moutonnées,* indicating that a thick body of ice once passed down the glen and ground out its sides and bottom. On the right hand the river Spean brawls and foams in a narrow ravine which it has cut for itself through the same hard schist. One could not find a better contrast of the results produced by the different denuding agents. Along the roadway all is rounded and smoothed as the ice left it; in the ravine below everything is angular and rugged; rains, springs, and frosts are there busy splitting up the schist along its numerous joints, and pushing large blocks of it into the river, where they are dashed against each other and slowly worn away.

Further up the glen the ice-worn aspect of the rocks becomes still more striking. When the observer reaches a point opposite the mouth of Glen Treig he finds that the striæ on the rocks, instead of running down the valley, actually mount the hill to the north, while the glen is cumbered with huge ramparts of glacier rubbish. It was this piece of scenery which so powerfully impressed Agassiz in his first visit to Scotland, and seemed to him such a demonstration of the existence of glaciers in this country. "I shall never forget," he says, "the impression experienced at the sight of the terraced mounds of blocks which occur at the mouth of the valley of Loch Treig, where it joins Glen Spean; it seemed to me as if I were looking at the numerous moraines of the neighbourhood of Tines in the valley of Chamounix."[1] Since Agassiz wrote, the district has been again examined in detail by Mr. T. F. Jamieson, to whose careful descriptions the reader should refer.[2] The glacier which came down Glen Treig went right across Glen Spean, and ascended the hills for some height on the opposite side. Mr. Jamieson has shown that it not only did so, but that it then branched into two, one part turning to the left down Glen Spean, the other taking to the right, through that part of the glen where Loch Laggan now lies, and striking thence into the valley of the Spey. The great size of the Treig glacier is further proved by the magnitude of its moraines. As it issued from its own valley into the wider strath of the Spean, it spread out in a fan-shape, and the piles of rubbish which it carried down on its surface were tumbled over along its edges. There they gathered into mounds, which followed the curving margin of the ice, and still stretch for some miles across the valley

---

[1] Agassiz, *On the Glacial Theory. Edin. New Phil. Journ.* xxxiii. 222.
[2] See *Quart. Journ. Geol. Soc.* xviii. and xix.

to mark where the glacier halted for a long while in its gradual decline. There are two chief lines of moraine, of which the outer one, a narrow, steep-sided mound, rises in some places sixty or seventy feet above its base. Its surface is loaded with fragments of gneiss and schist of all sizes up to blocks fourteen feet long. On the inner line of mound, which is often composed wholly of large blocks of syenite, Mr. Jamieson found one mass measuring twenty-six feet in length, and he compares the moraine to a ruined breakwater.[1]

The valley of Glen Treig furnishes everywhere proofs of the enormous erosion effected by the ice. Its sides, for a great height above the bottom, are shorn off, rounded, smoothed and striated, and it encloses a long lake, which points still more to the massiveness of the ancient glacier, for it lies in one of those basins of rock which, with Professor Ramsay, I believe have been excavated by ice.

Returning now to Glen Roy, the traveller should ascend that valley to see what light its famous "Parallel Roads" have to cast upon the history of the old glaciers of the Highlands. The same long straight line which, as he drew near to the Bridge of Roy, he noticed running high along the mountainside, on the south of the Spean Valley, is now seen to turn up Glen Roy, winding along the hills of that

[1] *Quart. Journ. Geol. Soc.* xix. 248.

valley with the same singular horizontality. When he gets several miles up the glen, he begins to see traces of two other terraces, until, on reaching a turn of the road, the long deep glen lies before him, with its three bars, straight and distinct as if they had been drawn with a ruler, yet winding into all the recesses of the steep slopes, and coming out again

FIG 14.—VIEW OF THE PARALLEL ROADS OF GLEN ROY FROM THE GAP.

over the projecting parts without ever deviating from their parallelism. The "roads," so long a subject of wonderment and legendary story among the Highlanders, and for so many years a source of sore perplexity among men of science, seem at last to be understood. Each of them is a shelf or terrace, cut by the shore waters of a lake that once filled Glen

Roy. The highest is of course the oldest, and those beneath it were formed in succession, as the waters of the lake were lowered. They are seen not only in Glen Roy. A little beyond, where the first good view of the glen is obtained, there is a hollow through the hills on the left side of the valley, marked on the maps as Gap. This hollow forms a short *col* between Glen Roy and a small valley that strikes away to the south-west. Standing on the top of the ridge, the observer looks up Glen Roy on the one side, and down this narrow valley on the other, and he can mark that, while the lowest of the parallel roads in Glen Roy runs along the hill-side, a short way below him the two upper roads come through the hollow, and wind westward into Glen Collarig. So that the old lake not only filled up Glen Roy, but also some of the other valleys to the west. Until Agassiz suggested the idea of a dam of glacier-ice, the great difficulty in the way of understanding how a lake could ever have filled these valleys, was the entire absence of any relic of the barrier that must have kept back the water. Mr. Jamieson has recently shown, however, that Agassiz's suggestion is fully borne out by the evidence of great glacial erosion, both in Glen Spean, and in the valley of the Caledonian Canal. The latter valley, as I have already pointed out, seems to have been filled to the brim with ice, which, choking up the mouths of Glen Gluoy and Glen Spean, served to pond back the waters of these glens. The Glen Treig glacier, in like manner, stretched right across Glen Spean and mounted its north bank. When the lake that must thus have filled Glen Roy and the neighbouring valleys was at its deepest, its surplus waters would escape from the head of Glen Roy down into Strath Spey, and at that time the uppermost beach or parallel road (1,140 feet above the present sea-level) was formed. The Glen Treig glacier then shrank a little, and the lake was thus lowered about eighty feet so as to form the middle terrace, which is 1,059 feet above the sea, the outflow being now by the head of Glen Glaster, and through Loch Laggan into the Spey. After the lake had remained for a time at that height, the Glen Treig glacier continued on the decline, and at last crept back out of Glen Spean. By this means the level of the lake was reduced to 847 feet above the sea, and the waters of Glen Roy joined those of Loch Laggan, forming one long winding lake, having its outflow, by what is now the head of Glen Spean, into Strath Spey. While this level was maintained, the lowest of the parallel roads of Glen Roy was formed. As the climate of the glacial period grew milder, however, the mass of ice which choked up the mouth of Glen Spean, and ponded back the water, gradually melted away. The drainage of Glen Roy, Glen Spean, and their tributary valleys was then no longer arrested, and as the lake crept step by step down the glen towards the sea, the streams one by one took their places in the channels, which they have been busy widening and deepening ever since. Such seems to have been the history of the mysterious parallel roads of Lochaber. Instead of tracing back their origin to the days of Fingal, they stand before us as the memorials of an infinitely vaster antiquity — the shores, as it were, of a phantom lake, that came into being with the growth of the glaciers, and vanished as these melted away.[1]

It would be far beyond the limits of this little work to describe or even enumerate the various valleys among the Highlands, where distinct traces are to be seen of the glaciers of the second period. It may be said, indeed, that there are probably few valleys, descending from the higher groups of mountains, where moraines and *roches moutonnées* are not to be seen. When we ascend a glen which receives the drainage of a connected cluster of lofty broad-bosomed hills, we may, with not a little confidence, expect to find, somewhere along its course, mounds of

---

[1] The above sketch of the successive stages in the history of the parallel roads is taken from Mr. Jamieson's memoir to which I would again refer the reader. See also the papers by Agassiz. *Proc. Geol. Soc.* iii. 327. *Edin. New Phil. Journ.* xxxiii. 217, and the *Atlantic Monthly* for June, 1864.

glacier-borne rubbish, or hummocks of ice-worn rock. The traveller who has occasion to pass northward, by the Inverness and Perth Railway, may see some good examples of the characteristic features of a glacier valley. As he ascends the course of the Garry beyond the Falls, bosses of hard quartz-rock and schist meet his eye, with their surfaces smoothed, polished, and striated, and heaped over with mounds of rubbish. The whole bottom of the valley, indeed, shows that it has been worn down by a glacier that descended from the mountains in front. The mounds increase in number towards the head of the glen, until, at the watershed, the ground from side to side is covered with vast piles of rubbish—true glacier-moraines. To the west lies Loch Garry—a lake seemingly held back by moraine matter, which is there cut into a succession of terraces, marking former levels of the water. The same piles of *débris* which can be traced to the head of Glen Garry cross the watershed, and go down Glen Truim, showing that the glacier of Loch Garry split upon the watershed, and sent one branch into Glen Garry, the other into Glen Truim. The deep pass of Drumouchter (1,450 feet above the sea), where this division took place, is as wild a scene as can be reached in the Highlands by a turnpike road—certainly by far the wildest through which any railway passes in this country. Even the comforts of a railway carriage and a good locomotive do not wholly deprive it of its terrors, but no one who cursorily makes its acquaintance under these circumstances can realize what the Pass of Drumouchter was in the old coaching days.

Let me refer to but one other locality where the relics of the second glacier period remain with a singularly picturesque vividness, and where some of their features can be better examined than in any other part of the kingdom with which I am acquainted. On the eastern coast of Sutherland the mountains come down to within a mile or two from the sea, the space between their base and the shore being occupied by a long narrow strip of comparatively level and cultivated ground. Of the glens which open upon this selvage of lowland, one of the largest is Strath Brora. It descends from the heart of the lofty Sutherlandshire mountains, and, after a course of some thirty miles, terminates at the inner edge of this narrow belt, nearly three miles from the sea. Looking at the map, one would be quite prepared to find glacier-moraines somewhere in this valley, but he could hardly expect to meet with such a striking series of mounds as roughen the dark plain between the mouth of the valley and the sea. The old glacier of Strath Brora must have spread out over the flat as soon as it escaped from its confinement between the walls of the glen, and the rubbish which it threw down gathered into long rampart-like mounds, that sweep round in a rude crescent form, from the foot of the hills towards the sea. Within the outer ridges, which are most continuous, there is a confused grouping of mounds, which, when they come together, enclose little pools of water, or basins of peat that mark where similar pools originally lay. Above the mouth of the valley the same detritus extends towards the loch—a quiet sheet of water lying under the shadow of dark crags, and held back still by the moraine rubbish of the glacier which once occupied its place. It is the relation of the moraine mounds to the sea-level, however, which gives them their greatest interest. They can be traced seawards in straggling hummocks and ridges, until, about a mile south of the village of Brora, one of the mounds is seen to overhang the beach, and a section of it, along with the gravelly beds on which it rests, has been laid bare by the waves. The glacier, in all likelihood, descended to the sea-level when these mounds were formed—an inference borne out by the loose materials on which they rest, and by the gravelly, water-worn character of the detritus of which the mounds are made up. And, if this be so, it shows us that so severe did the climate of Scotland still continue, that the valley-glaciers of some of the more mountainous districts continued to come down to the shore, until the land had risen to within forty or fifty feet of its present level.

At length the wide wintry mantle that had so long enveloped the mountains and valleys of the Highlands, crept from the sea slowly up the glens; each glacier shrank step by step backwards into the snow-fields of the uplands, and finally the snow-fields themselves melted away. The gradual increase of temperature, which thus drove the ice alike from land and sea, must have had no small influence upon the plants and animals of the time. Some of the forms which lived in the glacial period, such as the woolly elephant and the two-horned woolly rhinoceros, have become extinct; others, such as the reindeer and the musk-ox, have retreated to more congenial haunts in the far north. Among terrestrial plants we seem to see traces of the same northward migration of the more Alpine and Arctic species. They have been expelled from the warmer plains and lower hills, and have retreated into the mountains, on whose summits, struggling to maintain their place, they remain as the surviving relics of that northern vegetation which probably once covered the British Islands before they were cut off from the continent. The sea, too, furnishes proofs of the same gradual amelioration of the climate. In the deeper abysses of the western fiords, such as Loch Fyne and the Kyles of Skye, there are still lingering groups of the

Arctic shells which peopled our seas during the age of ice. Like the plants, they have been driven out by the migration of more temperate forms, and instead of now ranging from the shore-line down to the profoundest depths, they are confined to the latter parts of our seas, where they seem to be slowly but certainly dying out. The time may yet be distant, but it is probably not the less surely approaching, when the last of the Arctic forms, both of mountain-top and sea-bottom, will disappear, and when species of a more temperate character will spread over land and sea. And yet, such is the unceasing progress of terrestrial change, alike in organic and inorganic nature, that these newer forms will in all likelihood be themselves displaced by migrations from other parts of the globe, as the climate or the relative position of sea and land are changed, or as other mutations are brought about by those great geologic causes which, though seeming to operate at random, and wholly irrespective of either the animal or vegetable worlds, have yet been mysteriously linked with the grand onward march of life upon our globe.

CHAPTER VIII.

CAUSE OF THE LOCAL VARIETIES OF HIGHLAND SCENERY.

FROM what has been said in the foregoing pages, it appears that the larger elements of Highland Scenery—the grouping of the mountains, the excavation of the glens and straths, and the scooping-out of the lake-basins—are to be attributed to the action of the various powers of denudation—the sea, rains, springs, frosts, and moving ice. These features are everywhere to be seen, no matter what may be the nature of the rocks. We find the valleys cut out of gneiss, schist, slate, limestone, sandstone, quartz-rock or granite; but we feel sure there would still have been valleys and hills had the rocks been wholly different. The larger features of the scenery have probably been influenced rather by geological structure on the great scale, by anticlinal and synclinal axes, or lines of fault, or the boundaries of formations, than by mere diversities in the nature of the rocky masses that come to the surface.

But though the mineralogical changes in the Highland rocks may not have determined the framework of the country, they have had much to do in lending character to its surface. There is not a glen or strath where their influence may not be seen. It is this influence, due to differences of *weathering*, which has given to Highland scenery its variety, and to each district or to each rock its own peculiarites of outline. There are glens and valleys and lakes, for instance, in the Isle of Skye, around the Trosachs, and among the Cairn Gorm mountains. They all can be traced to the action of those powers of waste already described, their formation has been carried on in accordance with the same laws. Yet the results are in each case very different. And why? Because in each of the districts the rocks are distinct, and yield after a fashion of their own to the ceaseless attacks of time. In the north of Skye the valleys wind among soft green terraced hills of igneous rocks, and almost recall some of the pastoral uplands of the southern counties. Around the Trosachs the glens and lakes have been cut out of tough, gnarled schist, which is worn away unequally into knobs and bosses and steep craggy declivities. Among the Cairn Gorms the savage caldron-like corries and precipices have been carved out of granite—a rock which, from its usual decomposing character and its abundant vertical joints, combines in its decay a grandeur of lofty cliff with a smoothness of mountain-top such as none of the other Highland rocks can boast.

These local peculiarities of scenery are brought out only during the course of a long process of subaërial waste, when the rains, rivers, and frosts have had ample leisure for their quiet work. They must have reached their highest development just before the glacial period began. The ice-sheets, however, did much to break off the sharpness of the angles everywhere, and to give to the whole country a much tamer aspect than it had worn before. Since that time the atmospheric agencies of erosion have been busy upon the ice-moulded surface. They have been re-asserting their old sway, and though the track of the ice still remains singularly fresh, it bears everywhere the proof that it is disappearing, and that in time the rains and frosts will restore to the outlines of our hills and mountains all the ruggedness which they possessed before they were swathed in the wintry folds of the ancient glaciers. In comparing and contrasting, therefore, the different forms of scenery to which the different geological formations give rise, it should not be forgotten that the distinctions between them are not so great as they were once, nor so marked as they will be again, when the ice-worn surfaces have faded away.

Perhaps the most interesting way of tracing this relation of the minor outlines of the landscape to the nature of the rocks, will be to take some of the more important rock-masses of the Highlands and compare their scenery with each other. Beginning with the oldest formation of the British Islands—the Fundamental or Laurentian gneiss, we find it stretching as a broken belt along the western coast of Sutherland and Ross, from Loch Inchard for forty miles or more to the south. Nothing can well be more impressive in its barrenness than the aspect of this great fringe of gneiss. You stand on one of its higher eminences and look over a dreary expanse of verdureless rock, grey, cold, and bare, protruding from the heather in endless rounded crags and knolls, and dotted over with tarns and lochans, which, by their stillness, heighten the loneliness and solitude of the scene. Acres of sombre peat-moss mark the site of former lakes, and their dinginess and desolation form no inconspicuous features in the landscape.[1] Few contrasts of scenery in the Highlands, when once beheld, are likely to be better remembered than that between the cold grey hue and monotonous undulations of this ancient gneiss, and the colour and form of the sandstone mountains that rise along its inner margin. These heights are among the noblest in the

---

[1] If the hypersthene rock of the Cuchullin Hills of Skye belongs to this ancient formation, it forms an illustrious exception to the general monotony of outline above noticed. (See Plate II.) Several years ago I stated my belief that this rock was of metamorphic origin (*Trans. Roy. Soc. Edin.* vol. xxii. p. 633, *note*), and more recent observations of Dr. Haughton tend to confirm the inference. (*Geological Magazine* for February, 1865.)

Outline of Beinn Blabhein, Skye. (to shew the serrated form of a mountain of Hypersthene-rock)

whole Highlands. They consist of red Cambrian sandstone lying on the upturned edges of the gneiss, and with their strata so little inclined that these can be traced by the eye in long horizontal bars on the sides of the steeper declivities. Viewed from the sea, the gneiss belt runs in a line of bare rough hills and low headlands, among which, save here and there along a larger water-course, or on a straggling patch of gravelly soil, one looks in vain for tree or field or patch of green to relieve the sterility of these lonely shores. Behind, rise the sandstone mountains in a line of irregular but stately pyramids, their nearly level strata running along the hill-sides like lines of masonry. Here and there the hand of time has rent them into deep rifts, from which long mounds of rubbish are rolled into the plains below, as stones are loosened from the shivered walls of an ancient battlement. Down their sides, which have sometimes well-nigh the steepness of a wall, vegetation finds but scanty room along the projecting ledges of the sandstone beds, where the heath and grass and wild flowers cluster over the rock in straggling lines and tufts of green. And yet, though nearly as bare as the gneiss below them, these lofty mountains are far from presenting the same aspect of barrenness. The prevailing colour of their component strata gives them a warm red hue which, even at noon, contrasts strongly with the grey of the platform of older rock.

But it is at the close of day that the contrast is seen at its height. For then, when the sun is dipping beneath the distant Hebrides, and the shadows of night have already crept over the lower grounds, the gneiss, far as we can trace its corrugated outlines, is steeped in a cold blue tint that passes away in the distance into the grey haze of the evening, while the sandstone mountains, towering proudly out of the gathering twilight, catch on their giant sides the full flush of sunset. Their own warm hue is thus heightened by the mingling crimson and gold of the western sky, and their summits, wreathed perhaps with rosy mist, glow again, as if they were parts, not of the earth, but of the heaven above them. Watching their varying colours and the changes which the shifting light seems to work upon their strange forms, one might almost be tempted to believe that they are not mountains at all, but pyramids and lines of battlement—the work perhaps of some primeval Titan, who once held sway in the north.

These huge isolated mountains are probably the most striking memorial of denudation anywhere to be seen in the British Islands. Suilven, Quinaig, and other hills, rising more than 3000 feet above the sea, in Sutherland and Ross, are merely detached patches of a formation, not less than 7000 or 8000 feet thick, which once spread over the north-west of Scotland. The spaces between these mountains were once occupied by the same dull red sandstone; the horizontal beds of one hill, indeed, are plainly continuous with those of another, though perchance a deep and wide valley now lies between. The valleys have been worn down through the sandstone, and the fragments that have been left rise up into those strange pyramidal mountains that form so noble a feature in the

FIG. 15.—OUTLINE OF A MOUNTAIN BETWEEN LOCUS MAREE AND TORRIDON.—
(Dull red Cambrian Sandstone capped with white Quartz-rock.)

landscapes of the north-west Highlands. They stand there, like lonely sea-stacks in mid-ocean, and tell not less impressively of long ages of waste.

The Lower Silurian formation, which overlies the Cambrian sandstone, consists, at the base, of a thick group of white quartz-rocks with bands of limestone.

These strata, as I have already remarked, lie upon the red sandstone just described. They, too, have suffered sorely from denudation. They may be seen stealing up the backs of the Cambrian mountains, and even capping the very summits; and as they are marked by a snowy whiteness, the contrasting hues of the two rocks give rise to some of the most unexpected features in the scenery of these tracts. In certain phases of the sky, when the light falls brightly on the tops of these dark red pyramids, the white quartz-rock looks like a scalp of gleaming ice, and the long lines of white rubbish that seam the slopes might pass for glaciers which have shrunk up the mountains almost to the limit of perpetual snow.

Where a mountain consists wholly or mostly of quartz-rock it commonly tends to take a more or less conical form. Examples occur in Assynt, and further south between Loch Maree and Loch Torridon. The beautiful cone of Schehallien is an excellent illustration. But nowhere in Scotland can the whole of the distinctive features of quartz-rock scenery be seen on so grand a scale as among the mountains of Islay and Jura. In the latter island, the quartz-rock rises into the group of lofty cones known as the Paps of Jura, 2,569 feet above the sea which almost washes their base. The prevailing colour is grey, save here and there where a mass glistens white as if it were snow; and as the vegetation is exceedingly scanty, the character of the rock and its influence in the landscape can be seen to every advantage. The ascent of the mountains is impeded by a thick covering of loose angular rubbish, broken up by the rains and frosts from the rock underneath. But when once their summit is gained, the whole island and a wide panorama of sea and land beyond lies spread out as in a map below. Nothing can exceed the distinctness with which the lines of stratification in the quartz-rock are traced

FIG. 16.—OUTLINE OF SCHEHALLIEN (A MOUNTAIN OF QUARTZ-ROCK).

on the cliffs and along the ridges. We can follow almost the line of each separate bed of rock as it winds over hill and crag, valley and tarn, among solitudes that are haunted only by the red deer or the eagle. Here and there on the grey cliffs we detect the dark line of a basalt dyke pursuing the even tenor of its way towards the north-west, alike over precipitous mountain and deep glen. And far below, along the northern shore of that deep inlet which almost cuts the island into two, we can see the line of caves worn of old by the breakers out of the same pale rock, and the mounds of shingle that now lie between them and the sea.[1] There are few localities in the Western Islands where greater scope is offered to the painter than among the glens and corries of Jura. The scenery possesses in itself much of the rugged dignity of the Highlands; the mountains have the advantage of rising directly from the sea, and thus among scenes of the most lonely and savage wildness, there are glimpses of the wide Atlantic on the one side, and of the blue mountains of Argyleshire on the other. The island has not yet been inundated by the flood of summer tourists, and the artist may pursue his task undisturbed. He will find himself almost driven to enter upon a careful analysis of rocky scenery; for, amid the prevailing grey hue of the hills, his eye will be less apt to lose sight of the intricacies of form among the rich blendings of colour. And even if he should never make a picture out of his sketches, it will be strange if he does not find this enforced study of the structural character of rocks fraught with usefulness to him in all his after career.

[1] *Quart. Journ. Geol. Soc.* for May, 1861, p. 211.

The conical form of quartz mountains is likewise characteristic of other rocks which have a uniformity of texture. It seems to me to be due entirely to subaërial decay, and indeed to be a strong evidence of the reality and efficacy of this form of denudation. A homogeneous rock traversed with minute joints and cleavage lines, tends to break up into angular *débris* or shivers. Hence the thick coating of rubbish along the sides of the Paps of Jura, Ben Nevis, and hundreds of the other Highland hills. As soon as this detritus is broken from the rock it begins a slow descent to the valley below. The upper part of the mountain is thus left exposed to continual waste, while the sides, as they shelve downwards, are better and better protected under the coating of rubbish that becomes thicker towards the foot. When the structure of the rock and the activity of the powers of waste are duly proportioned, the result is that the mountain, worn away above and shielded under its ruins below, grows more and more tapering, until it passes into a perfect cone.

In this quartz-rock series lie some bands of limestone. These are easily recognised even from some distance, for they are commonly clothed with a bright verdure, which forms a pleasing contrast to the dun-coloured heath of the surrounding tracts. From a hill-top we may thus trace the course of a limestone band for miles. The rock itself only comes to the surface in knobs and hummocks, but its position is none the less surely marked by the line of soft green grass that winds from brown hillside into browner valley.

Over the quartz-rock and limestone comes that enormous and extensive group of gneisses and schistose rocks already noticed. These are not characterised by any uniform type of scenery—their variety of outline being doubtless traceable to their diversities of mineralogical texture, and to the consequently unequal waste which they undergo. The long chain of uplands that runs from the Sound of Mull to the Pentland Firth is formed of these gneisses and schists. Where the rocks are harder and more quartzose they give rise to a gnarled craggy outline, and though the mountains may not be lofty, yet their ruggedness gives them a wildness which almost compensates for their want of height. The scenery of the Trosachs is probably the most familiar example. But the same features may be seen on a much grander scale farther north, without that fairy-like garniture of mountain-ash, oak, birch, and willow, which lends such a charm to the Trosachs. Perhaps the defile of Glen Shiel in the west of Invernesshire, with its encircling group of lofty naked mountains, may be taken as one of the best examples of the more savage and rugged forms which these rocks assume. Dark masses of bare

Outline of Mountains on north side of Loch Hourn, opposite Barrisdil
(to show the spiry forms assumed by some of the higher eminences of the gneissose & schistose rocks. The lower rocks are ice-worn.)

rock seem there piled upon each other, giving a corrugated outline to the steep acclivities that rise up into an array of grey serrated ridges and deep corries, over which tower the peaks of Glenelg. Less accessible, but not less striking examples of the same savage scenery, may be found along the shores of Loch Hourn (Plate III.) and Loch Nevis. The height and the angular spiry forms of the mountain ridges, the steep and deeply rifted slopes, and the ruggedness and sterility of the whole landscape, distinguish these two sea-lochs from the rest of the fiords of the west coast.

Amid such scenes as these, the influence of the stratification and joints of the gneiss and schist on the decomposition of the rocks can be traced by a geological eye far along the summits and slopes of the mountains. To this influence are due those parallel clefts which give rise to dark rifts down the steep scarps, and to deep angular notches on the crests of the ridges. To the same cause also, combined with the unequal waste that arises from varieties in the texture of the rocks, we may ascribe that gnarled craggy contour so characteristic of the gneissose hills of the Highlands, as well as the frequent tendency of the summits to assume spiry forms. Sometimes a whole mountain has been worn into a conical shape, but more frequently it is along the crests or at the ends of ridges that this outline occurs, and the reason seems to be that the gneiss is usually too various in its texture and the rate of its decomposition to allow of the formation of a great cone like those of quartz-rock, while it is nevertheless uniform enough over lesser areas to give rise to small cones and spires along the summit of a mountain ridge.

Again, the influence of the internal structure of gneiss and mica-schist upon the rocky foregrounds of a Highland landscape must be familiar to many a visitor of the north. The mingling of mouldering knolls with rough angular rocks, the vertical rifts that gape on the face of crag and cliff as if they had been rent open by an earthquake, the strange twisted crumpled lines of the stratification, the blending of white bands of quartz with dark streaks of hornblende that vary the prevailing grey or brown or pink hue of the stone, the silvery sheen of the mica and the glance of the felspar or the garnets, the crusts of grey and yellow lichen or of green velvet-like moss, the tufts of fern and foxglove that rise above the clustered wild-flowers, the bushes of deep purple heather, and the trailing briars—these are features which we recognise at once as distinctively and characteristically Highland. With all deference I would urge upon our landscape painters the propriety of studying these details of rock scenery more than they have yet done. It is not as a mere mass

Geological structure and Landscape-painting. 221

of light and shade thrown into the foreground to give depth and distance to the picture that a group of rocks and boulders is faithfully rendered on canvass. There is an individuality even about the boulders, to which no conventional style of treatment can at all do justice. And no man will truly paint these features unless he is content patiently and lovingly to study them—a task which, I doubt not, he will find very pleasant in practice and eminently beneficial in result.[1]

[1] In the recent exhibition of the Royal Scottish Academy (1865) no feature struck me more than the conventionality of the Scottish artists in the painting of rocks. One of the more noted of their number is content to mottle the foreground of his Highland landscapes with lumps of umber and white, worked up indeed into the external outline of rocks and stones, but utterly without character. It seemed strange that in an exhibition containing so many pictures with Highland scenes as subjects, there should scarcely be one which showed that the painter had tried to study the individuality of rock-masses and boulders, over and above that of hills and mountains. In painful contrast to the work of the living artists was the picture of Pegwell Bay, by the late W. Dyce, R.A.—a work in which the geological structure of a chalk-cliff, not in itself a very attractive subject, was given with all the fidelity of a photograph, but with a poetry of tone and colour such as no photograph can give. The worn floor of chalk on the beach was admirably rendered, and the scattered stones were so well characterized that one could as easily pick out the large fractured flints from the rolled pieces of chalk as he could do on the beach itself. Yet this accuracy was not obtained at the expense of breadth and symmetry of treatment. It shows that, without descending to mere servile copying, a great addition of variety may be obtained for the foreground of a picture by a careful study of the structural forms of rocks. Mr. Ruskin's eloquent pages on this subject are familiar to every one. The most remarkable sketches of rock-scenery which I have

When the schistose rocks are of a softer and more uniform texture they form large lumpy hills with long smooth slopes covered with heath or peat, through which the rock seldom protrudes, save here and there where a knob of harder consistency comes to the surface, or where a mountain torrent has cut a ravine down the hill-side. Those wide tracts of the Highlands where the rocks are of this nature, possess a tame uniformity of outline which even their occasional great height hardly relieves. The traveller who crosses Ross-shire from Loch Broom to Dingwall, through the dreary Dirie More, will be able to realize this oppressive monotony, and to contrast it with the scarped and precipitous mountains that rise on the south round Loch Fannich. I do not know a better illustration of the effect of these softer schists in producing smooth-sloped hills than along the west side of the Firth of Clyde between the Kyles of Bute and the Gareloch. A band of clay-slate runs across the Island of Bute, skirts the firth by Innellan and Dunoon, crosses the mouth of Loch Long and the Gareloch, and strikes thence to Loch Lomond. It is easy to trace this strip of rock by the smooth undulating form of its hills, which remind us rather of the

---

have yet seen are those by my friend Mr. E. W. Cooke, R.A. In looking at them I am at a loss whether to wonder more at their scrupulous truth, or at the amount of thought and feeling which glows through each of them.

scenery of the southern uplands than of the Highlands. Behind the clay-slate lies a region of hard quartzose rocks, and the contrast between their rough craggy outlines and the tame features of the clay-slate is a familiar part of the scenery of the Clyde. It is to these harder rocks that we owe the ruggedness of the mountains that sweep from the shores of Loch Fyne through Cowal, across the Holy Loch, Loch Goil, Argyle's Bowling Green, and Loch Long, into the heights of Ben Lomond. Their craggy Highland character, when seen from the east, is not a little enhanced by the softly undulating contour of the clay-slate hills that come into the nearer landscape.

The granite (including the syenite) of the Highlands is not always characterized by a special type of scenery. Sometimes, as in the Moor of Rannoch, it covers leagues of ground without ever rising into a hill; or, as seen from the top of Cairngorm, it swells into wide, tame, undulating uplands; or it mounts in huge craggy precipices far up into the mists, and encloses dark tarns like Loch Aven, or it sweeps into dome-shaped eminences like the red hills of Skye, or into a stately cone like that of Goatfell. (Plate IV.) I have already remarked, that some of these various and apparently incongruous forms may be found combined in the same district, nay, even in the same mountain. From the summits of some, the granite mountains in the Grampian chain, the eye wanders over a wide, smooth, undulating table-land of hill-tops, and yet one or more of the flanks of each of these mountains may be a dizzy precipice 2,000 feet in descent, with its rifts of winter snow hidden deep from the sun. Such is the character of the highest parts of the Grampians,

>"Around the grizzly cliffs which guard
>The infant rills of Highland Dee."

Granite is usually traversed with innumerable joints, both parallel and oblique to each other, whereby the rock in weathering is broken up. The form which the granite mass assumes under the action of the wasting powers of nature depends greatly upon the angle of its surface. On a horizontal or gently inclined surface of granite, rains and frosts are comparatively helpless to split open the joints, for the upper layer of the rock breaks up into angular rubbish, which, though always wasting away, is always renewed in such a manner as to protect the rock below, and to preserve the uniformity of the surface. But where a vertical wall of granite rises into air, it may tend for a long while to maintain its precipitousness; for if it decays with tolerable equality throughout, slice after slice will be removed from its face, and if the springs, frosts, and streams at the foot of the cliff are active enough, the accumulated rubbish below may be swept away as it is loosened from the

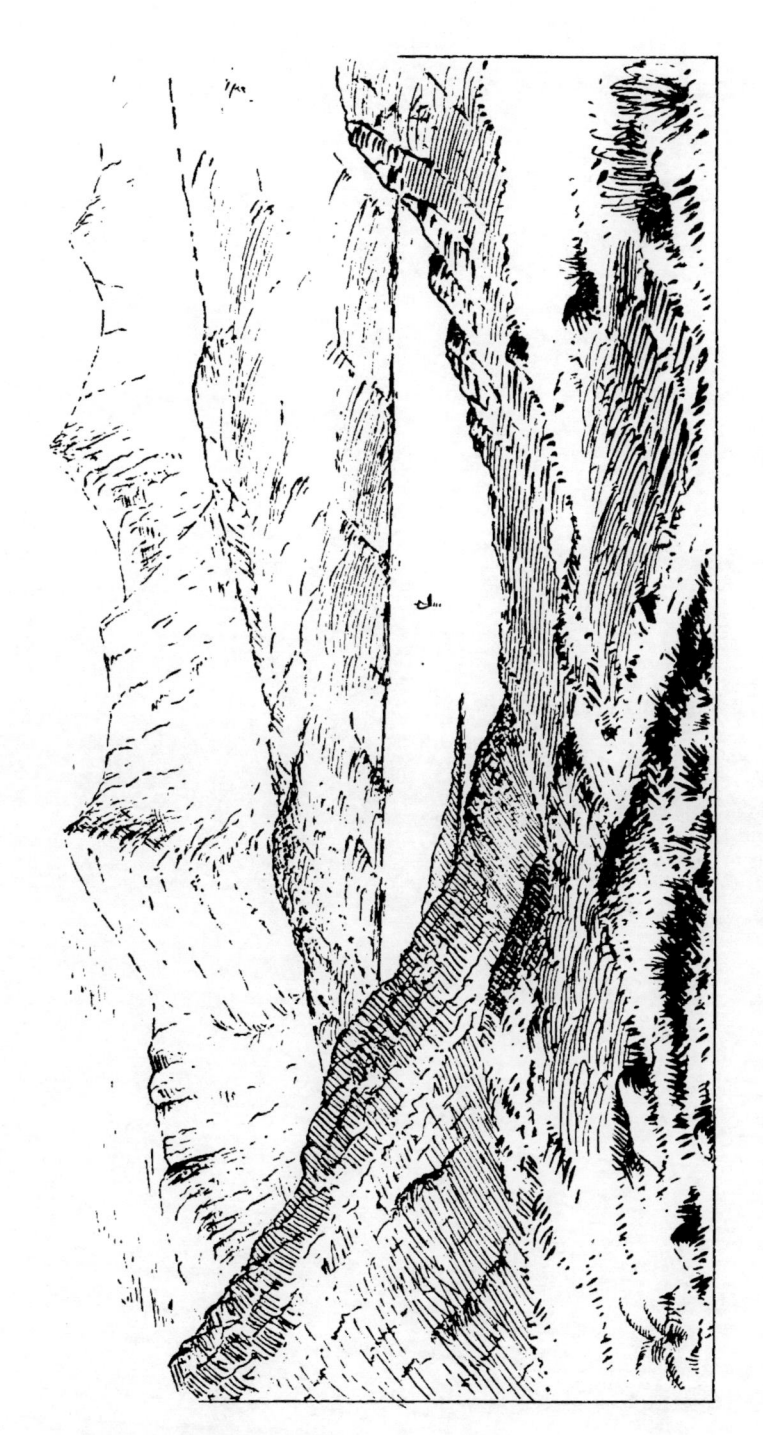

Outline of conical & dome shaped mountains of Syenite, Loch Ainort, Skye.

crags above. Hence the precipice would shrink backward into the mountain. To this unequal weathering I believe we should ascribe the singular extremes in the scenery of granite mountains, as well as the picturesque forms which are often assumed by groups of granite boulders.[1]

Throughout the Highlands, where the older palæozoic rocks adjoin formations of later date, there is usually a well marked contrast of scenery. Thus along the shores of the Moray Firth the brown rough mountains of the interior are fringed with a border of fertile ground, marking where the Old Red Sandstone takes the place of the schists. Further north a similar contrast shows where the Oolitic sandstones and shales of Sutherlandshire run as a strip along the coast, at the base of the line of rounded bare conglomerate hills which rise above them. On the west side of the island also, the Liassic and Oolitic strata, owing to the comparative richness of their soil and their low level, are sharply marked off from the Cambrian and Lower Silurian mountains which surround them.

[1] I must confess, however, that the Highland precipices present not a few difficulties to the geologist who would explain their origin and persistence. I should much like to know to what extent they and the deep glens from which they rise were modified during the glacial period. One may feel sure that they are not due to subterranean convulsion, though he may not be able to follow back to the first beginning the various stages of their formation during the slow excavation of the rock.

As a final example of the close relation between the contour of the Highland hills and the nature of the rocks that compose them, let me refer to the Inner Hebrides. These islands, from the north end of Skye to the south of Mull, are formed in great part, of greenstone, basalt and other masses of volcanic origin, piled above each other in horizontal or gently inclined beds, to a height of many hundred feet. Some of these rocks weather away more rapidly than others. Hence the profile of a hill in these districts

FIG. 17.—VIEW OF A TERRACED TRAP HILL, ISLE OF EIGG.

usually shews a succession of terraces, each of the harder beds standing out sharply against the sky like one of the steps of a staircase. Moreover, the general decay of the rocks gives rise to a rich loam, and thus these trap-islands are marked by the verdure of their valleys and the close bright sward which is found in patches even on the tops of their hills. No one who, in a sunny day, has sailed through the Sound of Mull can fail to remember how strikingly these features are there brought out. The green mountains of Mull rise from the water's edge, sometimes in steep rocky banks seamed with glistening waterfalls, and sweep inland, terrace above terrace and hill after hill, till their minor features are lost in the distance. The profile of one of these terraced slopes sometimes shows twenty or thirty distinct steps with gentle green slopes between them. The effect is often heightened where a soft shale or yellow sandstone is banded with the greenstones and basalts; for then, as the soft stratified intercalation moulders away, the igneous rock above it may be traced as a dark crag, running far along the side of the hill, and keeping parallel with the other terraces overhead and below. The landscapes in the interior of Mull and of the northern half of Skye owe the singularly artificial look of many of their lower hills, and the general terraced character of the whole surface, to the manner in which the bedded igneous rocks have yielded to denudation. The penning of these lines recalls to the writer many a pleasant reverie among the wilds of the Inner Hebrides, when, sitting in the light of the long autumn evenings, he used to mark the sinking beams strike along the sides of those truncated pyramidal hills, revealing terrace over terrace in alternate bars of dark crag and green slope—features that were but faintly seen in the glare of noonday—to look over the wild heathy uplands that stretched around to right and left in utter solitude and stillness; to watch how hill-top after hill-top would lose its blush of sunset, as if the dying day were slowly climbing the steps cut along the flanks of the western hills, and how the chill shadows, struggling upward from dark and lonely glens, crept up the same gigantic staircase until the whole landscape melted into grey gloom, and the night began to fall.

These and other local but characteristic parts of Highland scenery are to be traced, then, to the varying nature of the rocks of the country. I have already said that before the great erosion of the Ice Age began, these peculiarities were in all likelihood more strongly marked than they are now, and that, slowly recovering from the effects of the glaciation, they are returning to their former condition. I have been often struck with the progress of this change along the shores of Loch Fyne. The hard quartzose rocks opposite Tarbert are beautifully ice-worn and smoothed; their striæ, still fresh and clear, may be seen running out to sea under the waves. The lower parts have been protected from decay, owing partly to the recentness of their upheaval into dry land, and partly to their having been shielded by a coating of boulder-clay, not yet worn away from the bays. But above high-water mark, though the track of the old ice-sheet is still strikingly shewn, the rocks have begun to split up along their joints. Hence the low cliff that rises along the shore is rent into endless chinks and clefts, large angular blocks have been detached from it, and its base is cumbered with the ruins. Some of the islands show well the union of the glaciated outlines with this subsequent weathering. They still rise out of the water in long flat curves, like so many whales,— the form that was impressed upon them by the ice; yet they are split across along the joints into open cracks which one might fancifully compare to deep paralle gashes made across the whale's back.

FIG. 18.—ICE-WORN ISLETS IN LOCH FYNE, WEATHERING ALONG THE PARALLEL JOINTS OF THE ROCK.

The examples already cited from Ben Nevis and elsewhere, of the waste of cliffs and ravines, illustrate this obliteration of the marks of the ancient glaciers. It is unnecessary to multiply instances of a feature of Highland scenery which may be seen more or less distinctly on almost every hill-side and valley. Nor need I again point to the numerous ravines and river-channels which have been excavated or deepened since the ice disappeared, and which show that the same forces that carved out the valleys are still busy at their task.

Since the glacial period came to a close, not only have the various powers of waste been ceaselessly at work upon the land, there has likewise been an upheaval of the country to a height of fully forty feet above the level at which it stood when the glaciers crept back from the mouth of Glen Spean. The later stages of this rise will be further alluded to in a subsequent chapter in their relation to the kingdom generally. Forests, too, have sprung up and disappeared. Lakes have given place to bogs and peat-mosses. And man, a more rapid agent of change than the elements, has done much to alter the aspect of the Highlands. I think it better, however, to defer the notice of these later changes until the solid framework of the rest of the kingdom has been considered.[1]

---

[1] It will, of course, be understood that the scope of this volume permits me to treat only of those geological changes of which there are marked proofs in the scenery of the country. Hence I must pass over the evidence of oscillations of level afforded by the sunk forests, and other subjects which, though of great interest, do not specially elucidate the present inquiry.

CHAPTER IX.

THE SOUTHERN UPLANDS.

FROM the iron-bound coast of St. Abb's Head, on the one side of the island, to the cliffs of Portpatrick on the other, there stretches a continuous band of hilly ground, sometimes called the South Highlands, and which for the sake of clearness has been referred to in the foregoing Chapters as the Southern Uplands. This tract corresponds closely to the area occupied by the Lower Silurian formation, and its geology on the large scale is remarkable for simplicity. The whole district from sea to sea consists fundamentally of hard greywackè and shale, with occasional limestone bands, of Lower Silurian age, arranged in highly inclined or vertical strata that strike from south-west to north-east. These rocks, though hardened and even in some places changed into schist, serpentine, felstone, granite and other rocks, have not suffered from the same wide-spread metamorphism which has so altered the Lower Silurian rocks of the Highlands. Yet they have been squeezed into endless foldings, as if, to use the oft-quoted simile of Sir James Hall, they had been great piles of carpet, and had been crumpled up by a strong pressure from the north-west on the one side and from the south-east on the other. Hence the folds run from south-west to north-east— that is, at right angles to the direction of the pressure. And if the student would see the results of the process, let him traverse the rocky shores of Wigton and Kirkcudbright, or the cliffs of Berwickshire. He will there find a cross section of the rocks, and see the hard greywackè and shales bent into great arches and troughs, or squeezed into little puckerings, and will be able to trace these plications following each other from top to bottom of the cliffs, mile after mile along the coast. He will thus learn that in a long section of strata that are piled on end there may be endless repetitions of the same beds, very much as the leaves of a bound book are made up of numerous foldings of a comparatively small number of sheets; the top of the arches of the rocks having been cut away by denudation, somewhat as the edges of the leaves have been pared off by the book-binder.[1]

One result of such a visit to the coast sections of the Silurian uplands is to assure us that, just as in the

---

[1] See this structure shewn in Sections II. and III. along the margin of the accompanying map.

Highlands, so here, the arches of the strata have not given rise to hills, nor have their troughs formed valleys. Indeed, along the crest of the sea-cliffs we see the rocks cut sharply off by the surface of the country, whether they consist of hard greywackè or crumbling shale, whether they are on end or gently inclined, and whether they have been thrown into anticlinal or synclinal axes. It is plain that to whatever origin the present irregularities of the ground are to be assigned, they are not due to the upward and downward folds of the rocks. It will be seen in the sequel, that these uplands, like the Highlands, are a stupendous monument of denudation; that a vast thickness of rock has been ground away from their present surface, and that their hollows and hills have been determined by the same powers of waste that have played a like part in the history of the northern half of the kingdom.

By referring to the map the reader will observe that, in addition to these contorted strata, the Silurian tract of the south of Scotland contains a few outlying patches and strips of the later formations, which run up into the hills from the plains on either side. Thus the Old Red Sandstone lies in bays along the northern edge of the Lammermuirs, caps their summit south of Fala, and ascends from the low grounds of the Tweed up the valley of the Leader. The Carboniferous rocks indent the Silurian belt in a tongue which runs south from Cumnock, and contains the little coal basin of Sanquhar. The Permian breccias and sandstones likewise mottle the surface of the older rocks of the chain in strips and patches, in the valleys of the Annan and Nith, and even westward on the coasts of Wigton and Ayrshire. Wherever a large enough mass of such later deposits occurs it forms a distinct feature in the landscape.

With these trifling interruptions, the Silurian belt is found to have a well-marked scenery of its own, which is confined with notable strictness between the boundary lines of the Silurian rocks. Its north-western limit in particular is as well marked as the line that separates the Highlands from the Lowlands between the Clyde and Stonehaven. In East Lothian and Edinburghshire it mounts into the long chain of the Lammermuirs, whose steep bare front seems to project headland after headland into the fertile plains below, as a high river-bank rises out of its alluvial haugh, or as a lofty sea-cliff sweeps upward in promontory and bay above the waves that foam along its base. Across the counties of Peebles and Lanark, the edge of these uplands, though still well defined geologically, is sometimes not so marked in the landscape, owing to the rise of the ground in its front. But in Ayrshire it regains well-nigh all the boldness which marks its course through the eastern counties, and from the head of Nithsdale to the sea at Girvan, it is traceable in the long line of abrupt rocky hills which overlook the coal fields of Cumnock and Dalmellington, and rise so picturesquely out of the woods and corn-fields of the vale of the Girvan Water. The south-eastern border of the district is less exactly defined. There is, consequently, on that side an occasional gradation from the characteristic features of the great upland country into that mingling of wild moorland and cultivated valley which gives so peculiar a charm to the landscapes of the Borders.

The scenery of this broad belt of hilly ground is distinct from that of any other part of the kingdom. It maintains, indeed, a great uniformity and even monotony throughout its whole extent. No one can journey, however, through Galloway or Carrick, and thence through the uplands that stretch north-eastward to the German Ocean, without marking that this long chain of high grounds divides itself naturally into two not very unequal portions, each of which, while retaining the same family likeness, possesses nevertheless certain individual and distinguishing features of its own. The valley of the Nith passes completely across the Silurian region, and its course serves as an approximate boundary line between the two districts. This may be partly seen, even from a glance at the map; the tracts that lie between the Nith and St. Patrick's Channel will be observed to be broken up into irregular groups of mountains and hills dotted over with lakes and tarns, but traversed by few large rivers. The country between Nithsdale and St. Abb's Head, on the other hand, will be found to be marked as long connected chains of hills, nearly destitute of lakes, but with numerous confluent valleys, whose united waters, after a course of many miles, enter the sea as important rivers—the Clyde, Tweed, Esk, and Annan. But this distinction is, of course, much more marked in nature, where it can be seen in all its details.

The north-eastern half of the Silurian belt from Nithsdale to the German Ocean may be regarded as a wide undulating table-land, cut into coalescing ridges by a set of valleys which are usually narrow and deep. It has no determinate system of hill-ranges, the grouping of its eminences seeming in most cases to be defined by the circumstances which aided or retarded the excavation of the intervening hollows. Thus its seaward end, forming the heights of Lammermuir, when seen from the plains of East Lothian, has a long undulating summit, with an average level of 1,500 or 1,600 feet above the sea, and rises abruptly with a steep bare slope high above the rich champagne country that stretches to the shore. Standing on the north-western verge of these heights, on such an eminence for instance as Lammer Law (1,733 feet), the spectator sees below him, to the north and west, a rolling plain of woodlands and corn-fields, dotted with villages and mansions, down to the edge of the blue firth, and ranging westward beyond the crags and hills of Edinburgh. But he has only to turn round to the south and east to look over a dreary expanse of bare hill-top and bleak moor—wide lonely pastoral uplands, with scarce any further trace of human interference visible from this height, than here and there a sheep-drain or grey cairn. Far away south, beyond the limits of this solitary region, the Eildon Hills, Ruberslaw, and all the long chain of the Border heights eastward to the Cheviots, rise up with a softened outline from the green vale of Tweed. The surface of the Lammermuirs, like that of the greater part of these uplands, is singularly smooth. It is coated with short heath or coarse grass, save where a mantle of peat covers the hollows, or where the streams keep open their channels through the bare drift or hard rocks. So smooth, broad, and grassy are these hill-tops, that they may be traversed from Lammer Law to the eastern end of the chain without showing anywhere the solid rock at the surface, and but for the distant view of the rich lowlands lying far below, the traveller might walk mile after mile in the belief that he was passing over a piece of wild moorland, such as occurs in the lower parts of the country, instead of the summit of a chain of hills, some 1,500 or 1,600 feet above the sea. If, however, while moving, along the ridge he approached its edge, especially towards its western end, he would find that it descends abruptly into the plains, and is deeply trenched with gullies and narrow glens, through which its drainage escapes to the low grounds.

These heights of Lammermuir may be taken as a fair sample of the general scenery of the country between the German Ocean and the vale of the Nith. Yet in the higher parts of the district the smoothness and verdure of the hills are here and there exchanged for bold rocky scarps, bare crags and cliffs, and deep narrow defiles, that remind us now and then of parts of the Highlands. Such is the nature of the romantic pass of Dalveen among the Lowther Hills. Where the ground rises into the group of Broad Law and Hart Fell, the same features are seen in the deep dark glens that lead into Moffatdale,—as those of Black's Hope, Carreifran, and the Grey Mare's Tail; in the solitary glen of the Talla, and in the crescent-shaped cliffs of White Coomb and the Loch Craig, that "frown round dark Loch Skene."

Yet, though the flanks of the hills which form the higher parts of the uplands are thus cut into rocky declivities and narrow defiles, the prevailing character of a table-land is still impressively retained. Let the geologist ascend to the top of the Broad Law which, at a height of 2,754 feet, overlooks the whole of the surrounding country. The summit of this hill is so wide and level that, as has been suggested, it might be used as a race-course. Standing upon its flat heathy surface he sees around him a vast sweep of hills rising one after another, with long smooth summits that join on to each other and form, where seen from this height, a wide table-land. Were it not, indeed, for the deep valleys that can be traced threading their way through these hills, a stranger spirited away and set down on the Broad Law, might easily imagine he had been taken to some league-long moorland in the lowland plains. And even if he cannot get to the top, save by the prosaic and somewhat toilsome process of climbing, his walk, while it will give him a lively sense of the height of the table-land, will not diminish the wonder with which his eye wanders over the landscape, nor will it in aught lessen his conviction that this great expanse of elevated moorland must at one time have been an undulating plain, and that but for the scooping out of its valleys by subaërial waste, it would be a great plain still.

The scenery of these valleys is likewise characteristic. The heathy or peaty covering that often lies along the flat summits gives place to a coarse sward as the hills slope towards the streams, which they do sometimes steeply, sometimes gently, yet almost always with a smooth grassy surface, broken now and then by a scar or knob of grey rock. Where they approach each other closely they form such narrow deep glens as those of the Talla and Manor Waters, and where they are less shelving and wider apart, the broader space between their bases is spread out into alluvial meadow lands like those of the Tweed. Except along the water-courses, or where they have been planted along the lower parts of the hills, as a shelter for sheep, there are no trees. Nor do we meet with that union of crags and scars and broken ground, with masses of purple heather, ferns and wild flowers which enters so largely into a typical Highland landscape. It is in short, a smooth, green, pastoral country, cultivated along the larger valleys, with its hills left bare for sheep, yet showing enough of dark bushless moor to remind us of its altitude above the more fertile plains that bound it on the northern and southern sides.

Yet, with all this tameness and uniformity of outline, there is something irresistibly attractive in the green monotony of these lonely hills, with their never ending repetitions of the same pasture-covered slopes, sweeping down into the same narrow valleys, through which, amid strips of fairy-like meadow, the same clear stream seems ever to be murmuring on its way beside us. Save among the higher districts, there is nothing savage or rugged in the landscapes. Wandering through these uplands, we feel none of that oppressive awe which is called forth by the sterner features of the north. There is a tenderness in the landscape,

> " A grace of forest charms decayed
> And pastoral melancholy,"

that in place of subduing and overawing us, calls forth a sympathy which, though we cannot perchance tell why it should be given, we can hardly refuse to give. It may be, indeed, that with this feeling, human associations have much to do. For all this wide region of hill and valley is a part of that Border Country which has been hallowed by story and song. One cannot wander without dreamy thoughts of the past by Gala and Tweed, Ettrick and Yarrow, with their castles, and peels, and chapels, lonely and grey, and the traditions that seem to cling with a living power to every ruin and hill-side. And though, sharing in Wordsworth's experience, we may "see but not by sight alone," and allow "a ray of fancy" to mingle with all our seeing, we come back to these bare hills and quiet green valleys ever with fresh delight, and find that as we grow older they seem to grow greener, and to enter with a renewed sympathy into the musings of the hour.[1]

---

[1] To the reader who has not wandered through these uplands in sunshine and storm, I cannot hope to convey an adequate idea of their fascination. There is an interesting account of their different influence on Washington Irving and on Scott, in Lockhart's *Life of Scott*. See also Wordsworth's two exquisite poems of *Yarrow Unvisited* and

The south-western half of the Silurian region stretches from Nithsdale to Portpatrick. As I have already said, it is marked by the same great features as the north-eastern half between the Nith and St. Abb's Head. It has the same character of a wide table-land cut into distinct ridges by systems of valleys. Yet it shows a good many local peculiarities. It rises in Galloway into a cluster of mountains, of which the highest, Merrick, is 2764 feet above the sea — the most elevated ground in the south of Scotland.

These heights present the same Highland-like wildness as those to the north-east of Nithsdale, but on a far larger scale. Their tops indeed are smooth, but their sides are often deeply gashed with gullies and glens, or scarped with abrupt precipices. No scenery in the whole of the uplands of the southern counties can compare, for naked and rugged grandeur, with the glen's of the Merrick hills. There are no trees, no cultivation, no trace of man. The mountains descend in a wilderness of shattered crags into the dark glens, where the streamlets are often caught in little tarns, or dashed in foaming cataracts over ledge and cliff. Between these hills and the coast of Wigtonshire, lies a region of rough moor, mottled

Yarrow Visited. Descriptions of the scenery of these regions abound in Scott's novels, as, for instance, in *St. Ronan's Well*, *The Abbot,* and *The Bride of Lammermuir.*

FIG. 19.—OUTLINE OF THE TABLE-LAND OF THE SOUTHERN UPLANDS SEEN FROM ABOVE THE VILLAGE OF BARR. THE HIGHEST PEAK IS MERRICK, THE MOST ELEVATED POINT IN THE SOUTH OF SCOTLAND.

with endless hummocks of glacier drift, and scores of gleaming lakes.

Some of the scenic peculiarities of this region will be again alluded to in connexion with its old glaciers. Its greater wildness is perhaps to be attributed partly to the fact that it has undergone a far more decided metamorphism than any other part of the Silurian belt of the southern counties, partly to its greater height, and partly to the effects of the mass of ice which its elevation and extent enabled it to foster. The metamorphism is unequal and sporadic, but the general contour of the surface sometimes approaches closely to that of the metamorphosed rocks of the Highlands. Nevertheless, these broken rugged features are, at the best, but partial. They do not efface the traces of that undulating line which connects hill-top with hill-top in one wide sweep of table-land. Seen from the western side of the moors, or from the plains of Ayrshire, the glens and corries and defiles disappear, and we trace only a long mass of high ground, sloping gently away from a central point. (Fig. 19.)

THE SURFACE OF THE SOUTHERN UPLANDS, AN OLD SEA-BOTTOM.

In the wide, smooth, undulating surface of this great table-land of the southern counties the geologist can have little hesitation in tracing the results of a long protracted denudation by the sea. The lesson of enormous waste which we have seen to be taught by every Highland mountain top, is brought home to us not less vividly here. For it requires but a casual scrutiny to perceive that these long flat summits, instead of being made by the broad surfaces of horizontal strata, have been in reality planed down upon the upturned edges of contorted greywackès and shales. In crossing a smooth hill-top among these uplands, we pass over bed after bed, tilted on end, crumpled, inverted, broken ; yet the whole complex mass has been shorn away to a common level. By prolonging the truncated arches of the rocks (as shown in Sections II. and III. on the accompanying map), some idea may be formed of how vast a mass of material must have been worn away, and how entirely the surface of these high grounds has been fashioned by denudation. The cutting of such a great undulating plain out of hard rock must have been mainly the work of the sea, during perhaps many ups and downs of this part of the earth's crust. In short, the long belt of high ground between St. Abb's Head and St. Patrick's Channel is an ancient sea-bottom ; the broad green tops, dotted to-day with sheep and grouse and black-cock, took their levelled outline under the grinding power of the breakers, and partly, perhaps, of drift-ice borne by ocean-currents.

In what geological period did this wide-spread denudation begin? Within certain limits it is possible to answer this question. As the rocks are of Lower Silurian age, their denudation must be later than Lower Silurian times. They are overlaid here and there, especially in East Lothian and Berwickshire, with cakes of conglomerate and sandstone belonging to the Upper Old Red Sandstone. These deposits are only fragments of a once continuous sheet that spread over a large part of the Silurian region. They are made of the waste of the underlying rocks, and by their contents and position they show that before they were laid down, the contortion and metamorphism of the Silurian rocks had been completed. They lie upon the truncated worn edges of the older strata, like a series of books laid horizontally upon a lower set placed on end. It is plain, therefore, that a great part of the levelling of this wide Silurian table-land was effected before the conglomerates and sandstones were deposited. The work was in all likelihood done in great part during the Old Red Sandstone period, and even perhaps to some extent during the accumulation of the conglomerates of the Upper Silurian series. Thus the cutting down of the great Silurian table-lands of the Highlands and of the Southern Uplands probably took place during the lapse of the same vast geological period, viz. that of the Old Red Sandstone, although, counting by years, the Highland table-land may be thousands of centuries older than that of the south.[1]

After the Silurian rocks had been thus planed down into a broad undulating surface, they were carried beneath the waves, and as they sank, their own ruins were heaped upon them in piles of shingle and sheets of sand. These deposits are still visible, here and there, in patches of red conglomerate and sandstone. It remains to be ascertained to what extent the Silurian region was ever continuously buried under later deposits. That the Upper Old Red Sandstone covered a great part of its surface, both at its eastern and western end, can be satisfactorily proved. The remarkable faults which bring down the Carboniferous strata against the northern edge of the Silurian belt, raise a strong suspicion that these rocks once stretched completely across the Uplands, connecting the Carboniferous area of Central Scotland with that of Berwickshire and the north of England. Again, the patches of Permian breccias and conglomerate resting in valleys of the Silurian tracts, show either that the Old Red and Carboniferous Sandstones never existed there, or (which seems to me the more probable supposition)

[1] The removal of the Old Red Sandstone and Carboniferous rocks from the Southern Uplands may point to subsequent marine denudation, so that though the great table-land had its formation begun during Upper Silurian or Old Red Sandstone times, it may not have been ground down to its present level until a far later geological period.

that they had been cleared out of those valleys before the Permian deposits had been laid down.

FORMATION OF THE VALLEYS OF THE SOUTHERN UPLANDS.

To understand the progress of its subsequent changes, we must bear in mind that as the broad undulating plain, laboriously cut out of the Silurian rocks, sank stage by stage beneath the waves, it was covered over, at least in part, with a mass of detritus, hundreds or thousands of feet in thickness. When again this overlying mantle of later deposits, hardened into conglomerate, sandstone, shale, limestone and other rocks, rose into land, it had all to be worn away before the old Silurian table-land could again be laid bare. Since its first re-emergence it has doubtless been often sunk and raised anew. Each time that it came to the level of the breakers (or drift-ice), the cake of newer strata would suffer abrasion more or less extensive, according to the length of time it was exposed. So that with the sweeping away of the overlying Old Red and Carboniferous Sandstones the sea may have had not a little to do. At the same time, it must be granted that a great deal of the waste may have been carried on by subaërial agents. There is one part of the subaërial denudation which can be approximately measured—the excavation of the valleys, for after what has been said in the previous chapters there need be no further argument here to shew that these valleys are really hollows that have been scooped out of the solid rock, and that the erosion has probably been performed by running water and the other atmospheric powers of waste.

In no part of Scotland is the relation of valley to water-course more strikingly shown than in these Southern Uplands. The valleys are usually narrow and deep, and the streams either fill up the whole bottom, or wind from side to side in short curves, laving the base of the opposite hill-sides alternately. With the exception of the very ancient depressions in which flow the Leader, Nith, and Annan, there are no wide dales or straths. The valleys seem in each case proportioned to the size of their streams, and wind through the table-land with all the sinuosity characteristic of channels cut by running water. Seen from a height, indeed, this district looks like a kind of model of the drainage-system of a country; for the hills are so smooth and uniform in contour, that almost the only prominent varieties of outline are those marked out by the confluent lines of valley and water-course. Following the same subdivision which was used with reference to the Highlands, we may class the valleys of these uplands as *transverse* or stretching across the high grounds, and *longitudinal* or running along them.

By far the larger number of the main valleys

belong to the *transverse* series. Beginning at the east end, we find the Monynut, Whiteadder, and Leader among the Lammermuir hills. Farther west are the Gala, Eddleston and Lyne Waters, and beyond the wide basin of the Tweed lie the valleys of the Esk, Annan, Clyde, Nith, Dee, Doon, and Cree, with others of less note. As a rule, these streams rise close to the north-western edge of the table-land, and flow across to the low grounds on the south-east. The Leader, for example, takes its rise within two miles and a half from the northern base of the Lammermuirs, and runs thence southward to join the Tweed. The Dee, also, has its source not three miles from the edge of the table-land on the one side, but fully fifty miles from the opposite margin. There is no example of a river rising near the southern edge of the uplands and flowing across them to the north. But not only do most of the large streams begin close to the north-western flank of the Silurian belt; two of them, the Lyne Water and the river Nith, actually take their rise beyond the Silurian belt altogether, and flow completely across it. The Lyne has its source among the Pentland Hills, from which it descends into a broad plain between Linton and Romanno. It then strikes right into the Silurian hills, and joins the Tweed. The Nith takes a still more singular course. Its springs well out of the north flank of the Uplands, and the infant river descends into the Ayrshire Coalfield, as if to wind away towards the Firth of Clyde. Instead of turning in that direction, however, at New Cumnock it bends round to the south-east, and then entering the Silurian table-land, traverses its entire breadth, and merges into the Solway.

How these streams have been able to flow across a broad tract of hilly ground completely from side to side is a question which will need much additional inquiry for its satisfactory solution. In trying to see how it is to be answered we should bear in mind that some of these transverse valleys are of immense antiquity, and that they were begun when the contour of the country to the northward, forming now the central lowlands, may have been vastly different. The depression in which the Leader flows, known as Lauderdale, is as old as the Upper Old Red Sandstone, and may be much older. The gravel and sand which in that ancient geological period filled up the valley, have been in great part worn away, and the long hollow has become a valley again. Yet in the same chain of hills there is another depression of similar age, from which the sediment that filled it has not been removed. It stretches between Dunbar and Greenlaw, completely across the present chain of the Lammermuirs. The red conglomerate and sandstone still choke it up, and it remains a buried valley of as least as high antiquity as the Upper Old Red

Sandstone. Nithsdale, in like manner, cannot be of younger date than the Carboniferous rocks which lie along its bottom. Both Nithsdale and Annandale must have existed as deep transverse valleys when the Permian conglomerates and breccias were in the course of deposition. The depression filled by the waters of Loch Ryan and Luce Bay seems to be another transverse land-valley, once stretching away to the north-west, and as old as the Carboniferous period, for both Carboniferous and Permian strata line its western sides.

Again, it must not be forgotten that the long line of faults by which the greater part of the north-western margin of the uplands is bounded, shows that the ground to the north-west has been depressed. The table-land has, in fact, been broken through, and the part which remains was once continuous with a portion now buried deep beneath the lowland plains. How far it went northwards, whether it descended into a central valley, or sloped upward into still higher ground, there is, perhaps, no way of discovering. So far as we can tell, the Nith may originally have taken its rise on a higher part of the table-land, now sunk by the dislocations, and on the re-excavation of the valley, after the Carboniferous period, part of the drainage of the high-lying coal-measure ground to the north would naturally find its way into this old channel.

The *longitudinal* valleys are most numerous towards the eastern end of the table-land, the more important being those of the Teviot, Borthwick, Ettrick, and Yarrow, and that of the Tweed as far down as Peebles. They are also well-marked at the west end in the glens of App, Stincher, and Muck. The course of the Tweed is singularly suggestive of the relation of the one class of valleys to the other. From its source down to its junction with the Lyne, the Tweed flows north-eastward, though without strictly conforming to the north-easterly strike of the strata. It is turned round to the south-east by the Lyne, and after winding through the gorge of Neidpath, it is again bent to the south-east, where it receives the Eddleston Water. The united stream of the Ettrick and Yarrow has been able to deflect it once more for a few miles to the north-east, but as soon as it reaches the valley of the Gala, it immediately turns round to the south-east, and flows out into the Old Red Sandstone plains. We see here a sort of compromise between the two systems; yet, though by much the largest body of water comes down the longitudinal valleys from the south-west, the influence of the dominant (and, I believe, the older) class of valleys prevails, the drainage being bent round to the south-east, and carried in that direction out of the Uplands.

As the general strike of the Silurian rocks in this

part of the country runs from south-west to north-east, there can be little doubt that this structure has determined the course of the longitudinal valleys. The influence of the dip of the rocks, and of dislocations, in directing the excavating agents, is beautifully shown by the valleys in the south-west of Ayrshire. Glen App runs along the line of a fault.[1] The course of the Stincher for miles is parallel to but not coincident with a large fault, which may have given rise to some long hollow at the surface, whereof the river at first took advantage. But instead of rigidly following that line, the watercourse now lies some way on the north side of it, and actually cuts across it twice.

While the longitudinal valleys are to be traced to subaërial erosion effected under the guiding influence of the prevailing strike of the rocks, the transverse series, like those of the Highlands, are due to some cause apart from geological structure. They pass over all the rocks indiscriminately, and they show no proof of being cross fractures. They seem to me to have had their direction given them by the first seaward flow of the rain that fell on the table-land as it rose out of the waves. The ground probably

---

[1] Recently ascertained by Mr. James Geikie, in the course of the geological survey of Ayrshire. Professor Ramsay has detected the same fault on the west side of Loch Ryan, but there it does not give rise to any such marked feature at the surface as in Glen App.

had a gentle slope towards the south-east, and winding among the sea-worn channels, the drainage would arrange itself into water-courses, of which the main direction would be south-easterly. Once selected, these channels would be deepened and widened until, as age followed age, the upraised table-land was cut down into a network of valleys. The north-westerly slope of the table-land has been in great part thrown down by faults, and covered over with later formations.

It is unnecessary to enter into further details of the process of subaërial waste, for it has gone on in the same way in these high grounds as in the Highlands. Numerous examples might be cited of the recession of two glens towards each other and of the final result of this degradation in the formation of a valley or pass across a ridge. Nor need more than a passing allusion be made to the influence of the character of the rock upon the aspect of the scenery. I have already noticed the rugged outlines of part of the south-western half of the district, as contrasted with the smooth monotony of the north-eastern half, and have referred to the greater metamorphism and more unequal texture of the rocks of the Highland-like tracts as probably related closely to the difference of scenery. There is a wonderful uniformity throughout the whole Uplands in the nature of the rocks, and there is a corresponding uniformity in the character of the landscapes. Where a change takes place among the mineral masses, it sometimes shows itself, as among the wilds of Merrick, in deep corries and rugged glens, where the Silurian strata are highly altered and pierced with granite, or in a green fertile valley like those of Lauderdale and Nithsdale, where the older rocks are covered with Old Red Sandstone or some later formation.

INFLUENCE OF ANCIENT GLACIERS AND ICE-BERGS ON THE SCENERY OF THE SOUTHERN UPLANDS.

Although the present system of valleys among these high grounds has probably been marked out by the erosive action of running water, no one can wander among them without being struck with the proofs which they still retain of having been modified by the ice of the glacial period. Indeed, the whole surface of this part of the country bears that general smoothed flowing outline so characteristic of glacial action. Yet, owing to the tendency of the rocks to decay and to conceal themselves under a smooth coating of turf or heath, the polished and striated surfaces which tell so unequivocally of the movements of glacier or ice-floe, are comparatively rare. Now and then, where a recent landslip of boulder clay has taken place, the rock is laid bare with its striæ and scorings as fresh as when they were made. From the evidence of such occasional exposures, it appears that, like the Highlands, the Southern Uplands were deeply buried under ice which pressed its way down towards the low grounds outside. The heights about the sources of the Clyde, Tweed, Annan and Yarrow, formed a nucleus from which the ice crept down the diverging valleys. The great mass of high land in Galloway and Dumfriesshire became a still larger centre of movement. Even yet, the smoothed and grooved surface left by the ice-sheet as it overrode hill and valley may be traced for many miles away from the mountains, and the striæ are invariably found to point to the distant heights from which the glaciers moved outwards to the sea. Indeed on the south side of the high grounds the very trend of the minor valleys seems connected with the seaward divergence of the ice.

A further proof of the presence of the ice is found in the thick till or boulder-clay which lies in most of the valleys, and even mounts far up the hill-sides. This deposit, as I have already had occasion to remark, is so characteristically developed in the Lowland region that its description may be advantageously deferred till a succeeding chapter. The upper boulder-clay and the overlying sands and gravels, belonging to the period of submergence, likewise take their place among the superficial deposits of this district. But they, too, had better be included in the account of the similar accumulations of the Midland Valley.

There can be no doubt that after the high grounds of the south of Scotland were cased in ice, they began slowly to sink into the sea. How far the submergence went, whether it was total, or whether the higher hill-tops rose above the waves as dreary ice-clad islets, cannot be easily determined. The difficulty is the same as that already noticed with regard to the Highland mountains. When the country was upheaved again, the climate was still so severe that the ice continued to creep down to the sea, and in so doing it must of course have largely, if not wholly, swept away the traces left by the sea on the surface of the rising land. During the subsidence, also, there may have been much abrasion of the land by the grating of bergs and ice-rafts, which came partly, perhaps, from the north, and partly from the unsubmerged hills of the neighbourhood.

The records of this second series of glaciers remain still fresh and distinct. Among the valleys that run up into the mass of high ground between the upper part of Tweeddale and the sources of the Yarrow, the moraine mounds are as perfect as if the glaciers had only vanished a few years ago. In ascending the defile of the Talla, above the picturesque linns, we come upon mound after mound, sometimes fifty or sixty feet above the stream which has cut its way through them. They run in curves across the glen, from the foot of the hill on the one side to the corresponding slope on the other, the curves having their convex sides facing down the valley. Each mound marks of course a pause made by the glacier as it shrank, step by step, up into the narrowing snowfield at the head of the glen. Beyond the top of the Talla valley two deep semicircular recesses or corries have been scooped out of the sides of the mountains. One of these, that of the Midlaw Burn, is accurately described in Wordsworth's picture of a similar corry in the Cumbrian Chain.

> "A little lowly vale,
> A lowly vale, and yet uplifted high,
> Among the mountains;
> Urn-like it was in shape, deep as an urn,
> With rocks encompassed, save that to the south
> Was one small opening, where a heath-clad ridge
> Supplied a boundary, less abrupt and close."

A level meadow occupies its bottom, and the "heath-clad ridges" which close it on the south are successive

FIG. 20.—SECTIONAL VIEW OF LOCH SKENE, DUMFRIESSHIRE.

moraines—huge piles of rubbish cumbered with massive angular blocks of greywacké. The meadow was evidently at no distant day a lake, ponded back by the moraine heaps which have been cut through by the outflow of the water, until the lake has been drained. The other corry, or rather open recess, is that of Loch Skene. On the north and west sides, steep and, in part, precipitous slopes of craggy rock rise up from the margin of the lake, while on the east and south the water is held in by moraine mounds. I shall not soon forget the surprise with which, after climbing with my old friend and colleague, Dr. John Young, the ravine of the Grey Mare's Tail, and wandering through the heaps of glacier rubbish that lie along the valley above the linns, I saw from the top of the last moraine mound the deep blue waters of Loch Skene dancing in the sun. Everything around told of the old glaciers;—mound after mound stretching in crescent-shape across the valley, and coming down in irregular piles from the Midlaw Corry on the left, huge masses of rock still perilously poised on the summits of the ridges where they had been tumbled by the ice that bore them from yonder far dark cliffs, and then the lake itself so impressively the result of the damming back of the water by the bars of detritus thrown across the glen.[1] (Fig. 20.)

---

[1] Mr. Robert Chambers published an allusion to the moraines of Loch Skene in the second vol. of the *Edin. New. Phil. Journ.*, New Series, p. 184. The first description of them was, I believe, that given in the memoir of the *Glacial Drift of Scotland*, p. 160. Since that work was published the glacier relics of these high grounds have been carefully mapped, in great detail, by Dr. Young, and the chief results

But the mass of high ground between Nithsdale and St. Patrick's Channel was the chief seat of glaciers in the south of Scotland. Indeed the proofs of intense glacial action there are hardly less striking than in the mountainous parts of the Highlands. Most of the innumerable lakes of that district lie either in hollows among moraine mounds, or in ice-worn basins scooped out of the rock. Thus, between the foot of the range of the Merrick Hills and the Bay of Luce, the ground is one wide expanse of moor, roughened with thousands of heaps of glacier detritus, and dotted with scores of lakes enclosed among these rubbish mounds. The mass of ice which came down from the high grounds and moved westward and southward must have been very great. That it went out to sea, and hence that the moraine heaps of the great Galloway moors may have been to some extent submarine, is suggested by the composition of some of the mounds. If one continuous sheet of ice descended from the chain of mountains between the head of the Stincher and Wigtown Bay it must have been fully five and twenty miles broad. But whether in one mass, or in separate, yet adjoining, glaciers, the ice certainly moved towards the south and west with resistless force, for all the hills and hummocks of rock in its way are ground down,

are given by him in a paper communicated to the Geological Society. *Quart. Journ. Geol. Soc.* for 1864.

polished and striated, the boulder clay of the older ice-sheet and the marine drift are wholly, or almost wholly, removed over a space many square leagues in extent, and the surface of the country is strewed with mounds, ridges, and heaps of clay, gravel and boulders brought down from the mountains. It gives us a lively idea of the continued severity of the climate when we reflect that so limited a group of hills, the summits of which could scarcely have been more than 1,800 or 2,000 feet above the sea, should yet have nourished such masses of ice and snow. The seaward ends of the glaciers broke off as bergs, and carried huge masses of granite away north, south and west, dropping them over what are now the fertile fields of Ayr.[1]

As an example of extreme glaciation, I may refer to the valley of Loch Doon. A rugged amphitheatre, some four or five miles wide, lies among mountains that reach a height of fully 2,750 feet above the sea. The bottom of this great hollow is roughened with prominences which, were they not dwarfed by the lofty summits around them, would be looked on as noteworthy hills. It is dark with peat, shaggy with heather, and dotted with numerous tarns and lochs. Its collected drainage finds an exit by a narrow valley at the northern end, and it is there that Loch

---

[1] Part of this ground has been mapped in the course of the Geological Survey of Scotland by my colleague Mr. B. N. Peach and myself.

Head of Loch Doon, Ayrshire. (a rock basin; ice-worn rocks and glacier-borne boulders in foreground.)

up out of the lake. Not only so. For a hundred feet or more above the level of the water the rocks which rise above the end of the lake are similarly worn. There can be no doubt, I think, that the ice which filled the hollow of Loch Doon went up the slope at its northern end and so passed down into the valley beyond. The deep gorge of Glen Ness, by which the river escapes, seems to be partly the work of the glacial period, but much deepened since then by the wasting action of the roaring torrent which fills the narrow chasm from side to side. Besides the polished and striated rocks, the Loch Doon valley shows abundant moraine rubbish. Among the detritus, granite-boulders brought from the far mountains are especially numerous. They are sometimes thickly clustered in patches along the margin of the loch, or heaped on its islets. Some of the islets, indeed, look like the tops of moraine mounds appearing above the water. In short, there is no locality in the south of Scotland where the existence and effects of ancient glaciers can be more impressively seen, and none where, as it appears to me, the glacial origin of such rock-encircled lake-basins is more clearly evinced.

Since the last remnants of the great snow-fields and glaciers melted away from the uplands of the south of Scotland, there has been a good deal of minor change in the general features of the district. The crags and cliffs where the naked rock comes out into the light have suffered from long centuries of exposure to the elements, and their ruins are seen below them in streams of loose rubbish and piles of large blocks of stone. Many a runnel has deepened its first channel into a gully that runs as a narrow gash down the smoothed hill-side. The brooks and rivers too have been busy in eating away their banks and lowering their beds. Some of the most picturesque ravines, such as that of the Crichhope Linn in Nithsdale, have been cut by running water since the glacial period. The lower terraces and alluvial haughs that flank the margins of the larger streams, have likewise been made since then. Into the changes due to vegetation—the growth and disappearance of forests, and the formation of peat-moss—I do not here enter, reserving until a following chapter a brief reference to the nature and proof of such changes over the whole country.

CHAPTER X.

THE MIDLAND VALLEY.

GEOLOGICAL STRUCTURE.

BETWEEN the southern flank of the Highlands and the northern edge of the uplands of the pastoral counties, lies that wide hilly plain or valley which, for want of a better name, I have been accustomed to call the Midland Valley. It is only in the broad sense, as a band of lower ground between two mountainous tracts, that it can be spoken of as a valley; nor can a district so plentifully dotted with hills, and even traversed by long chains of heights, be in strictness termed a plain. Geologically, however, this belt of lowland country is a valley. Either side is bounded by Lower Silurian rocks, rising on the north in the rugged schists and gneisses of the Highland mountains, and on the south in the contorted greywackés and shales of the Southern Uplands. Between these two boundaries the rocks belong to newer formations, which may be broadly looked at as dipping away from the flanking hills, the oldest strata lying along the borders and the newest along the middle. This arrangement, however, has been sadly disturbed by numberless faults and depressions, so that some of the older rocks are found coming up to the surface far into the centre of the plain, while here and there the later rocks, in place being confined to the middle, stretch across to the margin and even go over into the Silurian district beyond.

Along the northern border of this valley there runs a broad well-defined belt of Lower Old Red Sandstone. It is exposed on the coast between Stonehaven and the Firth of Tay, whence it extends south-westward across the island to the Firth of Clyde. It contains a great abundance of igneous rocks—the product of volcanoes that were active in Scotland during the deposition of the Lower Old Red Sandstone. These volcanic materials form numerous detached eminences, as well as the long hill-ranges of the Sidlaws and the Ochils. There is no corresponding broad belt along the southern borders of the valley. The Lower Old Red Sandstone does appear there in a tract of moory heights between the Pentland Hills and the Ayrshire Coal-field, and has there been cut through in the well-known ravines and Falls of the Clyde. But it is surrounded by Carboniferous strata which overlie it and stretch away south to the Silurian uplands. The Upper Old Red

Sandstone follows the lower division for some way along the south flank of the Ochils, till it is overlapped by Carboniferous strata. It reappears from under these later deposits on the south side, rising into conspicuous hills among the Lammermuirs and appearing at intervals westwards into Lanarkshire.

The next formation in ascending series—the Carboniferous—covers by far the larger part of the central valley. From St. Andrew's Bay it stretches through Fife and Clackmannan to Stirling, and thence, sweeping round the flanks of the Campsie Hills, to the Clyde below Renfrew. A high bank of older igneous rocks here interrupts its continuity. On the west lies the large coal-field of Ayrshire; on the east the Carboniferous rocks spread over the valley of the Clyde up nearly as far as Lanark, and strike eastward through the broad lowland country till they abut against the Pentland chain. Immediately below the eastern slope of that range, they reappear and sweep through the rich lands of Edinburgh and Haddington to the sea at Dunbar.

It is the abundance of its igneous rocks which gives to the broad Midland Valley its most characteristic features. Besides the chains of the Sidlaws and Ochils, the Old Red Sandstone[1] contains the felspathic rocks of the Pentlands, and groups of hills in Carrick.

[1] Including both the Middle (?) series of the Pentlands and the Upper Old Red Sandstone.

At the top of the Upper Old Red Sandstone or base of the Carboniferous series there is a great development of volcanic material, which shows itself in the long chain of heights stretching from the Campsie Fells to the south of Arran, and from Greenock southeastward into the range of high grounds between the coal basin of the Clyde and that of Ayrshire. The lower half of the Carboniferous system up to the top of the Carboniferous Limestone abounds with igneous rocks. They form in some cases a small cluster of connected hills like those between Linlithgow and Bathgate, or like the Garlton Hills in Haddingtonshire. But their more usual form is in broken hummocky ground like that of the greater part of Fife, or in crags and small sharply-defined hills, such as those of Stirling and Edinburgh Castles, Binny Craig, the Bass Rock, North Berwick Law and many others. There are likewise igneous masses of still later age which have broken through even the newest rocks of the district, and, when of large enough dimensions, reveal themselves at the surface in the usual rugged features characteristic of the *trap-rocks*.

But for the existence of these igneous intercalations the stratified formations of the Lowlands would have formed mere wide undulating plains. It is the greenstones, basalts, felstones, and other igneous rocks which, rising up into bold hills, have relieved the uniformity of the surface, and have given rise to much of what we recognise as distinctly lowland scenery. It will be seen that the greater prominency of these rocks is a fresh and impressive proof of enormous denudation.

Allusion has already been made to certain long lines of fault which separate the formations of the Midland Valley from the Silurian rocks on either side. The boundary of the southern edge of the Highlands, and of the northern margin of the Southern Uplands, is in each case so remarkably straight that it naturally suggests a line of fault on either side of the lowland plain. There can be no doubt that the boundaries along part of their course are formed by dislocations. Thus the north-western flank of the Lammermuir Hills is defined by a fault which lets down the Carboniferous and Old Red Sandstones to spread over the low grounds of East Lothian. Another fault, with similar effect, runs along the foot of the Silurian hills that sweep eastward from Girvan. And across the centre of the island the Silurian boundary-line, which runs so evenly from sea to sea, seems to coincide with dislocations, perhaps older than the Carboniferous period. Along the edge of the Highlands the junction of the Old Red Sandstone with the metamorphic rocks appears sometimes to be defined by a fault, as at Callander. But in other parts, as from Bute along the Clyde shores to the Gareloch, and at Dunkeld, the conglomerates rest undisturbed on the edges of the schists, as sand and shingle may be seen resting against the base of cliffs and rocky shores, exposed to the dash of the waves. It may thus be that the Highland boundary was marked out before the deposition of the Old Red Sandstone, and that the edge of the Southern Uplands was in part, if not wholly, defined before the accumulation of the Carboniferous series.

Be this as it may, there can be no doubt that the broad lowland valley is geologically a valley of depression. It has sunk down below its ancient level, so that the strata which originally were deposited horizontally, or nearly so, are now bent into deep troughs. Thus, in the county of Edinburgh the coal-bearing beds lie in a long basin, the one lip or edge of which comes to the surface at Gilmerton, the other at Roman Camp Hill, the bottom of the basin being more than three thousand feet deep. If these beds could be flattened out it would perhaps be found that they must originally have lain high over the tops of the Lammermuir and Moorfoot Hills, and that they are thus merely the fragments of an extensive sheet of Carboniferous rocks that swept over the middle of Scotland, and may even have joined the coal-fields of England. That the different coal-basins of the Lowlands were once united is almost certain. At that time the strata must have been approximately horizontal, stretching across the country in one long, level sheet. Subsequent subterranean movements allowed certain portions to sink down into deep troughs, while the intervening parts were bent over into an arched form. During the lapse of centuries of denudation these connecting arches have been removed; the whole surface of the Lowlands has been worn down, and only those parts of the coal strata which have been preserved in the basins now remain.[1]

DENUDATION OF THE MIDLAND VALLEY.

From this slight sketch it is plain that the present comparatively level or undulating surface of the great lowland valley is not due to the horizontal arrangement of the rocks, nor merely to peculiarity of geological structure. A section across any part of the district shows that the strata, so far from being flat, are really curved to a remarkable degree. If no change had been wrought upon the surface after the rocks began to bend, we should now find them sinking and rising into broad folds, the bottom of the basins being sometimes several thousand feet below the tops of the arches. In all likelihood, however,

---

[1] The trough form of the strata filling up the Midland Valley is shown in the long sections on each side of the map. It will be seen from the truncated ends of the strata that the amount of denudation in this as in the other parts of Scotland has been great.

such irregularities were never suffered to make their appearance to this extent upon the surface. The subterranean movements by which the rocks were folded were probably very slow. Hence the arches, as they rose above the sea, would be wasted by the waves and the weather, and their detritus would be spread over the troughs that were sinking beneath the ocean. If the rate at which the rocks were curved and that at which they were wasted chanced to be equal, the upward folds would never make any show at the surface. And even if the arches were bent up somewhat faster than they could be worn away, the waste was perhaps always rapid enough to make the actual gain of land less in proportion than the total amount of upheaval. So much, at least, is certain, that if ever the folds of the strata rose into the air as wide dome-shaped hills and mountains, they have been all planed down as thoroughly as the furrows of a ploughed field are effaced by the harrow and roller. Millions upon millions of cubic yards of solid stone have thus been worn away, and the surface of the Lowlands has been again reduced to a general uniformity of level. Some idea of the magnitude of this denudation may be formed if we take one limited part of the district, and estimate the amount of rock removed from it. The Pentland Hills have at one time been covered with Carboniferous rocks to a depth of perhaps 6,000 or 7,000 feet. If the thickness of this covering be estimated at 5,280 feet, or one mile, and the hills as fourteen miles long by three miles broad, we find that the mass of material worn away must have been equal to forty-two cubic miles. This lost portion would surpass by more than five times the bulk of the present Pentland Hills; and, if it could be set down upon the Lowland Valley, it would form a group of mountains nearly a thousand feet higher than the loftiest of the Grampians.[1]

Again, the valley of the Firth of Tay lies on what is in geological structure not a depression, but an arch or anticlinal axis. Yet not only has the whole of the top of the arch been worn away, but the work of erosion has gone on until the upward curve of the rocks has been actually hollowed out into the present wide valley. The trappean ridges which along the edge of the Carse of Gowrie mount up in successive terraces, dipping away to the north-west, once rose high above what is now the Firth of Tay, and arched over till they came down into the Fife Hills. The connecting portion has been removed, but we see the truncated ends of the beds rising up to the south-east from the sides of the Sidlaws on the one hand, and up to the north-west from the Fife slopes on the other. The accompanying woodcut (fig. 21) is

[1] *Memoirs of the Geol. Survey. Geology of the Neighbourhood of Edinburgh*, p. 120.

FIG. 21.—VIEW OF THE FIRTH OF TAY FROM MONCRIEFF HILL, LOOKING EAST. (A Valley worn out of an anticlinal axis.)

from a sketch taken on the top of Moncreiff Hill, looking eastward towards the mouth of the Firth. It will be observed that the hills on the left, or north side, present their abrupt escarpments to the south, while those on the south side look in like manner to the opposite quarter, the sudden truncation of the beds pointing in either case most impressively to the extent of the waste.

Throughout the whole of the wide Lowlands, therefore, the present undulating surface is the result of enormous denudation. The abrupt solitary crags rising from smooth swept plains, the long, narrow valleys, the truncated ends of strata, as in Salisbury Crags, shooting up into mid-air, the deep defiles that trench the highest hills, even on the water-shed of the country,—all tell of the long-continued action of the abrading forces. Moreover, it requires but a casual survey of the rocks to see proofs of enormous dislocations, but these make no appearance at the surface. Fissures and faults which, had they not been effaced, would have left the country an impassable network of interlacing precipices and ravines, have been all planed down, until not even a geologist would, from the form of the ground, suspect their existence. In short, it is impossible to turn in any direction without meeting proofs of this vast and wide-spread denudation.[1]

[1] *Id. loc. cit.*

It will be observed that the chief hills and eminences throughout the Lowland Valley are formed of igneous rocks. There is a vague, and not uncommon, opinion that these rocks have been upheaved bodily through the other formations into the light of day, and that it is to the forms assumed by them on their eruption that the present irregularities in the surface are to be traced. Such an idea, however, cannot be held by one who knows that the greater part of the igneous material was ejected contemporaneously with the deposition of the strata among which it occurs, and that it has suffered along with these strata in all the long series of foldings and denudations. Even those igneous masses, which have been thrust up among rocks previously formed, can often be shown to have consolidated beneath the surface, and to be now at last exposed, only because the rocks under which they lay have been worn away. The truth is that the hills and crags of trap-rock are among the most striking proofs of denudation. For in by far the larger number of cases it may be shown that they were once buried deep under stratified formations now removed, and that they rise up into prominent elevations, mainly because they are harder than the sedimentary rocks, and better able to hold their own in the long warfare with the elements. Binny Craig, Arthur's Seat, North Berwick Law, Stirling Rock, the Abbey Craig, and many other isolated rocks throughout the Midland Valley, may be cited as illustrations; while of the larger and longer chains of hills conspicuous examples are seen in the Brown Carrick Hills of Ayrshire, in the range of heights which separates the coal-basin of the Clyde from that to the west, and runs north-eastward into the Kilpatrick and Campsie Fells, in the Linlithgow, Pentland, Garlton, and Fifeshire Hills, and in the chains of the Ochils and the Sidlaws. The reader who wishes to understand the structure of some of these trappean masses may consult the works cited below.[1]

Instead of describing, by way of illustration, the details of one of these examples, I shall give an account of the structure of the Scuir of Eigg.[2] The same lesson of vast waste is there taught which comes before us everywhere in the Lowlands; but in this case the details present a more impressive picture of the successive stages and final result of denudation than can perhaps be found in any Low-

---

[1] Maclaren's *Geology of Fife and the Lothians*. *Memoirs of the Geological Survey, Geology of the Neighbourhood of Edinburgh* (especially chapter xiii). On the Chronology of the Trap-rocks of Scotland. *Trans. Roy. Soc. Edin.* xxii. 633. (See *post*, p. 295, Fig. 22.)

[2] The above description is the result of a careful survey of the island, made by me in company with my friends the Rev. Mr. Macbride, of Ardmory, Bute, and Dr. John Young of the Geological Survey. Eigg has been several times described, but never before surveyed in detail. The singularly instructive history of the Scuir has thus hitherto escaped observation. The full description of the island will form the subject of a paper by Dr. Young and myself.

land hill. No feature of the western coast takes the traveller so much by surprise as the singular ridge which runs along the hill-tops at the south end of the island of Eigg, and forms the well-known Scuir. Seen from the sea on the east side, this ridge rises as a lofty massive column, towering to the height of some four hundred feet above the high ground on which it stands, and 1,335 feet above the sea. Its sides are quite vertical, so much so, that if one has a steady enough head to stoop over the edge of the precipice he may see its base four hundred feet below. What seems a broad and lofty tower, when looked at from the east, is really the abrupt end of a long narrow ridge, which widens out westward until it loses itself in a mass of rugged ground, abounding in little rock-basins filled with water. The Scuir itself, with these broken heights into which it merges, consists of the long-known pitchstone-porphyry (*c* in section on the map) almost everywhere columnar, the columns being sometimes piled up horizontally with their weathered ends exposed, sometimes slanting inwards or outwards, like a *chevaux de frise*, and often built round a hummock of rock very much in the way the peasants stack their peats.[1] In walking among these rocks one is constantly reminded of

[1] The drawing by Fortis of some columnar basalt in the Vicentin might have been taken from the sides of the Scuir of Eigg. See Lyell's *Elements*, 6th Edit. p. 612.

the fantastic arrangement of the columns in lava streams. And in truth this hard rugged rock, forming the highest part of Eigg, is the remnant of an old lava-flow. It lies upon a bed of compacted shingle, in which there is an abundance of coniferous wood, in chips and broken branches, yet so well preserved that, when newly taken out and still damp, it might be taken, but for its weight, for the relics of some old pine-forest buried under a peat-bog. The pitchstone-lava, however, has been a better preserver than even a peat-bog would have been. The shingle, as shown in the section on the map, lies on a worn surface of a much older volcanic series, consisting of almost horizontal beds (*a*) of greenstone and basalt, which, after being piled up sheet after sheet, until their accumulated mass had reached a thickness of many hundred feet, were hollowed out, apparently by running water, into winding channels. These water-courses can in part be traced under the pitchstone, with their gravel and drift-wood still remaining. At that time the island of Eigg must have been joined to some higher land, probably to the west or north-west, for the stream brought down with it blocks of hard Cambrian sandstone—a rock not found in Eigg, but abundant on the opposite island of Rum. What is now the ridge of the Scuir was therefore a valley watered by a stream that flowed with considerable volume to carry along the blocks, sometimes two or three feet in diameter, which are found in its shingle. It was when this was the aspect of the ground that the pitchstone was erupted. As a stream of molten-rock it rolled along the river-channel, and ascended for a short way the courses of some tributary streams, burying the whole under a mass of solid rock. Since that time the various powers of waste have been ceaselessly at work. They have succeeded in wearing down the land that united the greenstones and basalts to mountains of Cambrian sandstone, and they have made Eigg an island. But more than all, they have worn away the greenstones and basalts forming the higher grounds that bounded the old river-valley, and reduced them to slopes that shelve down into the sea; and so complete has been the change, that the buried valley, under protection of the singularly indestructible pitchstone, now runs along the top of a group of hills. What were once hills have disappeared, and what used to be a valley is now the crest of a lofty hill. The pitchstone which, when it rolled down that ancient water-course, sought the lowest level it could find, rises to-day into one of the most conspicuous landmarks on the west of Scotland. Yet even this firm rock is itself mouldering into ruins, for its fragments are strewn thickly along the base of the cliffs. Already it has been largely diminished from its original size, and we can look forward to a time when it shall have wholly disappeared.

The denudation of the Lowland Valley has probably to a large extent been the work of the sea, during many upheavals and depressions throughout a long course of geological periods. Viewed indeed on the large scale, this district may be regarded as a great plain of marine denudation, which has been modified in its outlines partly by geological structure, and partly by subsequent subaërial waste. The successive stages of this long course of denudation have still to be made out. At present we have only glimpses of some of them.[1]

The river systems of the Lowlands, and the subaërial denudation to which they appear to be mainly due, deserve careful study.[2] There are, in this part of

---

[1] Among the reasons why the plain of marine denudation in the Midland tract has so much more varied an outline than that of the Southern Uplands seem to be the facts that it lies upon many widely different kinds of rock, and that its site has during a number of geological periods been marked by great volcanic activity, and by many local, but important changes of level.

[2] I do not wish to undervalue the effect of unequal subsidence or elevation in producing wide depressions, and thereby influencing the form and grouping of basins of drainage both in the Lowland Valley and in the high grounds on either side. That such movements have taken place on a great scale, a slight acquaintance with the structure of the coal-fields is enough to show. Nevertheless, it cannot be denied that the surface of the country has been so planed down that its present inequalities bear but a slight proportion, and often have no relation at all, to the curvatures of the rocks underneath. The prominent

the country, three chief rivers, the Tay, Forth, and Clyde, the two former descending from the high grounds of the Highlands, and the latter from those of the southern counties. The basins of these streams are not separated from each other by continuous ranges of high grounds. Sometimes the watershed coincides with a chain of trappean hills, such as that between the Firths of Tay and Forth, but between the Clyde and Forth the ground undulates across the great coalfields from the Campsie to the Pentland Hills, where the parting of the waters may frequently be crossed without being noticed.

What may have been the form of the ground when

ridges and hills are composed of hard materials, the lower grounds as a whole lie on the softer rocks—a difference which must result, I think, from unequal denudation, and not from unequal movements of the earth's crust. For it can hardly be believed that only the harder rocks were upraised, or only the softer ones depressed. This is a subject, however, which requires very cautious handling. All that I can venture to say about it at present is that, while proofs of denudation stare me in the face at every turn, I have been unable hitherto to find any positive evidence that the origin of the valleys of the Lowlands has been connected with subterranean movements. These valleys sometimes run along synclinal troughs of this strata, but they likewiss cross lines of anticlinal axis, and in either case in such a capricious way as to show their independence of geological structure. Yet though the ancient curvatures of the strata have not had a main share in guiding the course of the valleys, there may have been subsequent local risings and sinkings of the land, of which no geological evidence remains, but which were not without their influence in the growth of the valleys. And though the chief work may still have fallen to the share of the denuding agents, we must not overlook the possible modifications produced by subterranean movements.

the drainage of the high lands to the north and south first began to flow across the present rocks of the Midland Valley, is a question to which no definite answer can as yet be given. We cannot tell what inequalities, due in part perhaps to unequal subsidence or elevation, and in part also to unequal marine denudation, may have existed to guide the streams in their seaward course. The rivers, however, have certain features in common, which may perhaps give some clue to the ancient configuration of the district. It is to be noted that they each break through the chain of hills at the head of their firths. The Tay has cut its channel through the hard trap-rocks and conglomerates that form the Sidlaw Hills, and flows across them in a deep valley between Kinnoul Hill and the Hill of Moncrieff. The Forth at Stirling has forced its way between the Ochils and the Campsies. The Clyde has severed the Kilpatrick Hills on the north side from those of Renfrewshire on the south, and flows in a broad stream between them. It may be that in one or all of these cases the breach of the hills corresponds with a line of dislocation. But we have no reason to believe that, if such dislocations ever existed, they were visible at the surface in the shape of wide open rents, through which a river could make its way. According to the analogy of all the faults and fissures which have been explored in pits and mines, they were probably close, their walls either touching each other or separated only by a few feet of mineral matter. But I am not aware that, save in the case of the Forth, there is any proof of dislocation at all. The defiles, or valleys, are comparable with hundreds of others in all parts of the country. If then the chain of the Sidlaws once ran unbroken to the south-west, through the Hills of Kinnoul and Moncrieff, into the range of the Ochils, of which geologically it is a prolongation, how could the Tay trench it? Two explanations may be suggested. The ridge of hills may have been cut across to a certain depth by two streams eating their way back towards each other, in the manner already pointed out as at work to the north of the Cromarty Firth. The transverse hollow thus formed would become a strait during a subsidence of the land, and might be considerably widened and deepened by the sawing action of the waves, until after repeated changes of level the trench was cut down so low that the drainage of the country to the north was drawn into it. Or we may regard the channel as dating from that immensely remote period when the present broad valley between the Sidlaw range and the Grampians was filled up with Old Red Sandstone, long since removed, and when the Tay flowed in a gentle depression worn by the sea across the trappean rocks. The river gradually worked its way downward through these rocks, while the whole surface of the country was undergoing denudation, both from atmospheric causes and from the sea. During the frequent submergences of the land, the breach would suffer from the waves as well as from subaërial waste, so that on each re-elevation, the river would tend to return to the gap through which it found a ready passage to the sea beyond. In what way soever we account for this defile we must rely, I think, on the help of the various powers of denudation, not on subterranean convulsion. And thus the high antiquity even of the minor features of the existing surface of the country is powerfully impressed upon the mind.[1]

The narrow valley at Stirling, through which the Forth flows out into the wide level carse lands, may possibly be connected with the prolongation of the large faults that bring down the Alloa coal-field against the southern flank of the Ochil Hills. If the line of weakness indicated by these faults ran westward it would pass between Stirling and the Bridge of Allan, and might well have shown itself at the surface as a long hollow worn down by springs and rains. And were this hollow once formed we can see how it might be dug out by the ceaseless wear and tear of the elements, until it became a channel fit for an ample river. When we cast our eye over this singular strait,

---

[1] The occurrence of the Upper Old Red Sandstone at Clashbinnie seems to indicate that the broad valley of the Firth of Tay is at least as old as that formation.

with its abrupt sides, and its flat meadow-like bottom stretching away into the wide mosses of Menteith on the one side, and the broad Carse of Falkirk on the other, we cannot but feel that in the cutting down and widening of this gap the Forth has had powerful helpmates in the waves of the sea. Yonder, indeed, not many miles to the east, is the blue firth; the level valley is plainly an old sea bottom, and sea-shells are dug up abundantly from its sands and clays. Even now so little is it above high-water mark that a depression of ten feet would send the tide up the valley for eighteen miles,[1] and if the land were sunk very little more, the firths of Clyde and Forth would meet, and a set of vexed tides would ebb and flow across the centre of Scotland. Such has doubtless often been the condition of the country in the geological past. Hence the deepening and widening of the strait at Stirling, like that of the Tay at Perth, may be traced in no small measure to the prolonged action of waves and tidal currents.

In the case of the Clyde at Bowling there is a transverse valley cut through the trappean rocks and sandstones. It resembles that of the Tay in its geological structure, and is susceptible of the same explanations which were suggested as to the origin of the defile below Perth.

While the three main rivers resemble each other

[1] *Trans. Roy. Soc. Edin.* iii. 268.

in thus breaking through a chain of hills to find their way into their firths, they present many points of difference in their respective courses across the Lowland Valley. Perhaps the most interesting is the Clyde. Drawing its waters from the very centre of the Southern Uplands, it flows transverse to the strike of the Silurian strata until, entering upon the rocks of the lowlands at Roberton, it turns to the northeast along a broad valley that skirts the base of Tinto. If the reader will glance at the map he will notice that at that part of its course the Clyde approaches within seven miles of the Tweed. Between the two streams, of course, lies the watershed of the country, the drainage flowing on the one side into the Atlantic, and on the other into the North Sea. Yet, instead of a ridge or hill, the space between the rivers is the broad flat valley of Biggar, so little above the level of the Clyde that it would not cost much labour to send that river across into the Tweed. Indeed, some trouble is necessary to keep the former stream from eating through the loose sandy deposits that line the valley, and finding its way over into Tweeddale. That it once took that course, thus entering the sea at Berwick instead of at Dumbarton, is probable, and if some of the gravel mounds at Thankerton could be reunited, it would do so again.[1] Allusion has already

[1] The occurrence of salmon in the Clyde above the Falls has been explained from the relative levels of the streams in the Biggar Valley.

been made to this singular part of the watershed.[1] Its origin is probably traceable to the recession of two valleys, and to the subsequent widening of the breach by atmospheric waste and the sea.

From the western margin of the Biggar flat the Clyde turns to the north-west, flowing across a series of igneous rocks belonging to the Old Red Sandstone series. Its valley is there wide, and the ground rises gently on either side into low undulating hills. But after bending back upon itself and receiving the Douglas Water, its banks begin to rise more steeply, until the river leaps over the linn at Bonnington into the long, narrow, and deep gorge in which the well-known Falls

"It is a singular circumstance," says Stoddart, in his *Angler's Companion for Scotland*, "that salmon and their fry have occasionally been taken in the upper parts of the Clyde above its loftiest fall, which, being eighty feet in height, it is utterly impossible for fish of any kind to surmount. The fact is accounted for in this way. After passing Tinto Hill, the bed of the Clyde approaches to a level with that of the Biggar Water, which is close at hand, and discharges itself into the Tweed. On the occasion of a large flood, the two streams become connected, and the Clyde actually pours a portion of its waters into one of the tributaries of the Tweed, which is accessible to, and frequented in the winter season by, salmon." *Angler's Companion*, p. 417. Yarrel states the highest salmon leaps to be from eight to ten feet; Stoddart supposes they may sometimes be more than twelve feet, and he says that in the Tummel the fish must leap eighteen feet, for they are caught above the falls. (But perhaps salmon ova might be carried by birds.) The Biggar flat, however, is not the only place in that neighbourhood where the watershed of the country crosses a nearly level valley. A few miles to the north the upper waters of the Tarth and the Medwin flow along the same meadow land, but the former stream turns eastward into the Tweed, and the latter westward into the Clyde.

[1] *Ante*, p. 148, note.

are contained. That this defile has not been rent open by the concussion of an earthquake, but is really the work of subaërial denudation, may be ascertained by tracing the unbroken beds of Lower Old Red Sandstone from side to side. Indeed, one could not choose a better place in which to study the process of waste, for he can examine the effects of rains, springs, and frosts, in loosening the sandstone by means of the hundreds of joints that traverse the face of the long cliffs, and he can likewise follow in all their detail the results of the constant wear and tear of the brown river that keeps ever tumbling and foaming down the ravine. A little below the town of Lanark, the Mouse Water enters the Clyde through the dark narrow chasm beneath the Cartland Crags. There, too, though

"It seems some mountain rent and riven,
A channel for the stream has given,"

yet after all it is the stream itself that has done the work. Nay, it would even appear that this singularly deep gorge has been in great measure, if not wholly, cut out since the end of the Age of Ice, for there is an old channel close to it, filled up with drift, but through which the stream has evidently at one time flowed.[1]

Running still in a narrow valley, the Clyde, after receiving the Mouse Water, hurries westward to throw

[1] See *Memoir on the Glacial Drift of Scotland*, p. 51.

itself over the last of its linns at Stonebyres, and to toil in a long dark gorge until, as it leaves the Old Red Sandstone, its valley gradually opens out, and it then enters the great Lanarkshire coalfield. From the top of the highest Fall to the foot of the lowest is a distance of rather more than $3\frac{1}{2}$ miles, in which the river descends about 230 feet, or 65 feet in a mile. From the Stonebyres Linn to the sea at Dumbarton the course of the Clyde is fully 50 miles, yet its fall is only 170 feet, or about 3·4 feet in a mile.[1] As it winds among its broad meadows and fair woodlands, no one ignorant of the geology of the district would be likely to imagine that this wide, level valley really overlies a set of strata which have been tilted up and broken by innumerable dislocations. Yet such is the fact. The flat haughs of the Clyde were not laid out until after the curved and fractured coal-measures had been planed down, and no external trace of these underground disturbances remained. The sea may have had much of the earlier part of the work to do, and may have lent its aid now and again during the successive uprisings and sinkings of the land, but we shall, perhaps, not greatly err in attributing mainly to the prolonged action of rains and frosts, and of the Clyde itself, the excavation of the broad valley in which the river flows across the coalfield until it reaches the sea between the hills of Renfrew and Dumbarton.

[1] See Petermann, *Edin. New Phil. Journ.* xlvii. p. 309.

CHAPTER XI.

CAUSE OF LOCAL VARIETIES OF LOWLAND SCENERY.

IN following the details of that vast denudation which has levelled the Lowlands, it is not uninteresting to trace, as in the case of the Highlands, how each marked variety of rock has not been worn away without imparting its own peculiar outlines to the scenery. Of course by much the most noticeable rocks in this respect are those of igneous origin—the felstones, greenstones, basalts, and trap-tuffs which cover so large a space and enter so conspicuously into the Lowland landscapes. I do not know of any large mass of such material that does not form a ridge or hill; even the smaller protrusions usually show themselves at the surface as little mounds, or craggy knobs. Moreover, the leading varieties of the igneous rocks weather each in a fashion of its own. Thus the felstones are often worn into smooth conical eminences, usually coated with turf which, when broken here and there along the slopes, allows long streams of angular rubbish to crumble from the rock, and slide down the hill. The noblest example in the whole of the district is Tinto, a huge mass of bright flesh-coloured felstone which rises to a height of nearly 1,700 feet above the Clyde, which washes its base, and 2,300 feet above the sea. Smaller and less perfect cones may be traced towards the north-east, from Quothquan into the Pentland Hills, and far to the east they reappear in Traprain and North Berwick Law. The cones are sometimes united to each other, thus forming a long range of hills, such as the Pentlands. In other cases the felspathic rocks have been worn into smooth undulating billowy hills, which nevertheless tend, ever and anon, to take the conical form, as may be seen in the eastern half of the Ochils. It is hardly necessary to remark, that though the cones sometimes look not unlike volcanos, as North Berwick Law, for instance, their form is entirely due to denudation. The reason that they take this outline seems to be the same as that which I have already suggested with regard to quartz-rock mountains. The felspathic igneous rocks have often a remarkable uniformity of texture; and, instead of breaking up into large blocks, crumble into loose angular shivers, which slide down and form a smooth covering for the lower parts of the hill, leaving the upper slopes exposed to continual waste, so that, as the hill moulders away, it becomes more conical. In the end, however, the cones must begin to get blunted at top, and as the angle of the declivity changes, the rate and tendency of the waste will be proportionately affected.

The greenstones and basalts on the other hand (and sometimes the felstones), are marked by a more craggy and rugged outline. Where they occupy a considerable breadth of surface, they rise into broken irregular ground, protruding ever and anon through the soil in rounded hummocks of dun-coloured rock, which crumbles into rich brown loam. Hence they present a curious blending of bright green sward with bare lichen-crusted stone. The numerous parallel joints in these rocks afford pathways for rain, springs, and frosts. The texture of the masses, too, varies indefinitely; sometimes we see the stone so firm and compact, that on its bare weather-beaten surface the dark augite crystals glance in the sun, while in other places it is decayed into a mere loose yellow or brown sand. Between these two extremes every variety of hardness may be seen, and hence probably the hummocky aspect of the ground over which greenstone and basalt prevail. Where one of these rocks has been intruded of old into the Carboniferous sandstones and shales, it may be seen rising into a steep-fronted crag, such as Stirling Rock, Binny Craig, or the rock on which Edinburgh Castle stands. The soft strata which surrounded and covered it have been all worn away, and the hard trap-rock now towers above the low grounds. It is to this denudation that the well-known form of hill called *Crag and Tail* is due. The worn edge of the bed of greenstone or basalt shoots up into a *crag*, and the strata which rest upon it, though swept off the top of the eminence, are often preserved on the declivity behind, and form in this way a *tail*.

FIG. 22.—SECTION OF THE FORM OF HILL IN THE LOWLANDS OF SCOTLAND KNOWN AS "CRAG AND TAIL."

a. "Crag" of basalt. *b.* "Tail" of softer rocks which have been worn away from around the hard basalt. *c.* Hollow often found in front of the crag. *d.* Covering of drift.

Around Edinburgh, where these features were first noticed, the crag usually faces the west, as in Salisbury Crags, Calton Hill, and Castle Rock, because the escarpment or outcrop of the rocks is in that direction, and consequently any rock which stood out above the others would naturally give rise to an eminence, steep on the west side and shelving on the east. Where the dip of the strata is to the west, and therefore their escarpment to the east, the crag will look eastward—of which there is an instance in Corstorphine Hill.[1]

[1] See the original descriptions of Sir James Hall, *Trans. Roy. Soc. Edin.* vol. vii ; Maclaren, *Geol. Lothians*, pp. 52, 218. It seems to me that "Crag and Tail" is a form of hill due mainly to the influence of the

If, instead of being inclined, the rocks are flat or nearly so, we find that the hard greenstone or basalt has served in some measure to protect the strata underneath. The hard rock then lies as a cake or capping on the top, while the sandstones or shales form the main mass of the hill. In the West Lomond of Fife, for example, the stratified deposits rise almost horizontally to a height of 900 feet above the vale of Eden, their bared edges being cut away into a steep declivity. Over them comes a thick bed of greenstone, which runs as a dark precipice along the crest of the slope, and supports a little patch of sandstone and limestone, on which lies the knob of greenstone that forms the conical summit 1,713 feet above the sea. It is plain that a mass of rock, fully 900 feet thick, has been hollowed out of the valley of the Eden, and, as the strata are flat, and show here and there their yellow edges along the green slopes, with the vertical bar of greenstone at top, the hill presents a suggestive monument of denudation. The

---

the geological structure of the rocks upon their denudation, though probably modified to some extent by the various agents at work during the glacial period. (See *Memoir on the Glacial Drift of Scotland*, p. 30.) But I am now much inclined to doubt whether any inference as to a general denuding agency acting from a given direction can justly be drawn from "crag and tail," as developed in the Lowlands of Scotland. This form of ground has been compared, and even identified with the *lee seite* and *stoss seite* of Scandinavian geologists; but there is really no analogy between them, for it can be shown that in the Scottish examples the crags or lee sides face the wrong way.

View of Loch Leven Kinrosshire, from above Crook of Devon, (shewing the denudation of the Lomond Hills).

water, and that after long ages of change, they have come at last to be shaped into ranges of quiet pastoral hills! In fig. 21 the influence of the bedded structure of the trappean rocks is seen in the escarpments of the Sidlaw range on the north side of the Tay, and in the Fife hills on the south.

It is thus the igneous rocks which, in the general waste of the surface, have, by their greater permanence, given rise to the hills and crags of the Lowlands. The stratified rocks as a whole are singularly featureless. They form the tame level groundwork which is relieved by the igneous masses. Now and then they rise into such rounded eminences as the conglomerate and sandstone hills of the Pentland chain, or the rolling moorlands that lie on the confines of Lanark and Ayr. But they are usually buried under a smooth-swept covering of drift, so that any features which they might have lent to the landscape are concealed. It is for the most part only in such confined spaces as ravines, or along the sea-margin, that we can learn after what fashion the softer rocks yield to the attacks of time.

The disappearance of the older stratified deposits under a wide-spread mantle of drift, leads us to consider how far the scenery of the Lowlands has been modified by the glaciers and icebergs of the long Ice-Age.

TRACES OF GLACIERS AND ICEBERGS IN THE LOWLAND VALLEY.

When the great ice-sheet of the glacial period began to settle down upon Scotland the main features of the broad Midland Valley were probably very much what they are still. There has been not a little obliteration of the minor details of the scenery; the hills have been rounded and smoothed, many an old valley and river-course has been partially or wholly filled up with boulder-clay. The whole country has been smothered up, as it were in drift, and hundreds of new hillocks and mounds have been scattered over its surface. Yet the larger elements of the landscape have in all likelihood undergone no marked change.

Throughout the Lowlands there is no chain of hills which seems to have been high and broad enough to nourish an independent group of glaciers. But the great ice-sheets from the Highlands on the one side and from the Southern Uplands on the other streamed down into the low grounds and through them to the sea. Hence the hills have that worn or "glaciated" surface so characteristic of Scottish scenery. And they still retain abundantly on their sides and summits the striation and moulding which mark the direction in which the ice moved. From the position of these markings we learn that the massive ice of the great Highland area came down into Strathmore and kept steadily southward in such force as to mount over the chain of the Sidlaws, and even, it would seem, over the Ochils, until it went out to sea by the basin of the Forth. Further west a huge body of ice descended from the north into the basin of the Clyde, filling the firth and overriding the hills on either side. From the Southern Uplands, the ice went north into the plains of Ayrshire, until, meeting the stream from the Highlands, it turned southward into St. Patrick's Channel. No wonder therefore that there should be *roches moutonnées* even on the tops of many of the Lowland ridges, and that the rocks should show such wide-spread proofs of having had their roughness rounded off.[1] But the hills of the broad Midland Valley had their sides and tops

---

[1] The evidence for the above statements will be found in the *Memoir on the Glacial Drift* already cited, as far as it had been ascertained by geologists up to the end of 1862. Since that time I have been able to extend the proofs of glaciation over a wide area in Ayrshire and the west of Scotland. I may likewise take this opportunity of stating that an upper boulder-clay which I had been disposed to regard as a mere local variety of the Till now seems to me an interrupted yet wide-spread deposit. The meaning of this upper clay began to dawn upon me in the autumn of 1863, when I mapped it out in detail in the south-west of Ayrshire. My colleagues in the Geological Survey likewise commenced to separate, where practicable, the upper from the under boulder-clay, and this method has been carried on since the spring of 1864. The reader will find a brief account of the two clays in a little volume published in October last, under the title of *Outlines of the Geology of the British Isles*—a handbook to accompany the geological map in Mr. Keith Johnston's Educational Series. See also *Proceedings of Royal Society of Edinburgh*, for February 6th, 1865.

smoothed and worn, not only by the downward movement of the great glaciers, but probably in some degree also by the icebergs, which, during the submergence of the land, broke off from the ends of the glaciers, or which were drifted from a more distant region. If, as the country settled down beneath the sea, a fleet of bergs were driven annually by the prevailing winds and currents over a particular tract of submerged land, the sunken hill-tops that were brushed and rasped year after year by those ponderous masses, would doubtless suffer no slight loss. They would be still further worn down, scratched and polished, for although drifting icebergs could not, as it seems to me, produce that universal striation which mounts over the hills and crosses the valleys, moulding itself to all the inequalities of the rocky surface, they might have done much in rubbing away and furrowing the slopes and summits along or across which they were driven.

When the surface of the country was thus ground down both by glacier-ice and floating bergs, a large amount of detritus must necessarily have been produced. It is this material which, over so much of the Lowlands, hides the rocks from sight. It is known as *till* or *boulder-clay*, and may be divided into an upper and an under part, each of which points to different aspects of Central Scotland during the glacial period. The lower boulder-clay or true till may be seen in the course of almost every Lowland brook, from the sea shores up to a height of perhaps 1,700 feet. Its thickness in the Carse of Stirling is said by Mr. Bald to be 160 feet, though possibly part of this mass may belong to the upper clay. The streams in the lower parts of the country have in many cases cut deep ravines through it, and it then stands up in steep walls on which the characteristic features of the deposit are well displayed. Certainly there are not many localities better fitted to perplex and discomfit an eager geologist than one of these cliffs of boulder-clay. He sees before him a stiff sandy clay, without any traces of stratification, full of stones of every size up to blocks several feet in diameter. These are grouped in no order whatever; large boulders and small pebbles are scattered indiscriminately through the clay from top to bottom. They are stuck at every angle, their smoothed and polished surfaces are covered with ruts and striæ running chiefly along the longer diameter of the stones, and if the face of the rock below be uncovered, it may be seen to retain the same markings. Using his hammer upon them he finds them to consist, almost wholly, of fragments from the rocks of the immediate neighbourhood. In a coal-measure district, for instance, he sees a mixture of bits of sandstone, shale, ironstone, coal and other Carboniferous strata, with a few pieces of the harder rocks of an adjacent geological area. He can perceive that this deposit must have been produced in the district from which it obtained its pebbles and boulders, but the mode of its formation has been for at least half a century one of the obscurest problems in Scottish Geology. From the almost universal striation of the boulders, and their local origin, it is now inferred that the boulder-clay or till has been ground up by a moving mass of land-ice, and has been deposited partly on the land, where it could remain protected from destruction by the moving glaciers, and partly in the sea under the sheet of ice that was pushed out from the shores and broke up into wandering bergs. (See Fig. 4.) It thus corroborates the inferences to be drawn from the striated rock-surfaces, and though there are still difficulties connected with the details of the process of formation and deposit, these might perhaps be easily and satisfactorily solved if a competent observer could bring himself to spend some time along the margin of the great ice-sheets of Greenland.[1]

[1] That arctic bergs probably had nothing to do with the formation of this deposit is indicated by the absence of far-travelled stones in the clay. After years of patient exploration, I have never succeeded in detecting in the Scottish boulder-clay a single stone which might not have come from rocks not many miles away. I have been in the habit of taking percentages of the stones at different localities in a district, and the result has invariably been to establish the prevailing local character of the deposit, and to point out the direction from which the ice moved. In the course of a ramble from Berwick-on-Tweed to the mouth of the Humber, Professor Ramsay and myself did not meet with traces of Scandi-

The Upper Boulder-clay is usually a looser and more gravelly deposit than the Lower, though the line of demarcation between them is not always very distinct. It contains larger boulders, and among them a greater per centage of stones that have come from a distance. It seems to be a thoroughly marine deposit, formed some little way from the shore, as the land was sinking, when masses of ice floated off, not merely from the neighbouring coast, but even from distant shores, and dropped their burthens of mud, sand, and stones upon the bottom. Marine shells, usually in a fragmentary state, occur in this deposit.

The boulder-clay furnishes us with some scanty indications of the denizens of the country during these wintry ages. Bones of the "Mammoth," reindeer, and of, probably, several species of oxen and deer have been from time to time dug up. And doubtless there must still remain the relics of other animals yet to be discovered in the older glacial deposits of Scotland, for scanty though the fauna could not fail to be, it need not perhaps have been less than that of Greenland. Besides the remains of mammalia, the Scottish till near Airdrie has yielded traces of land vegetation: thin beds of peat and trunks of

---

Scandinavian rocks till we had reached the mouth of the Tees, and they became more and more common the further south we went, as if the main trend of the arctic bergs had been towards the eastern or south-eastern coast of England.

trees in what seem to have been little lakes or tarns lying in hollows of the glacial detritus. And on a ridge overlying the valley where these interesting relics occur, shells of an Arctic type have been found, lying below a mass of boulder-clay, and serving further to support the inference that the boulder-clay was formed during a depression of the land, when successive terrestrial surfaces were submerged and covered over with marine accumulations.

Among the deposits which date from this same period of subsidence are certain beds of brick-clay, which have a classic interest in Geology. They are well seen along the low grounds on the banks of the Clyde below Glasgow, and on the shores of many of the sheltered bays and sea-lochs of the Firth of Clyde. They occur also along the eastern shore, at intervals, from the Forth to the Moray Firth. Their chief interest to the geologist arises from the fact that they contain an abundant series of shells, from which strong additional evidence is obtained as to the former intensity of the climate. More than a quarter of a century has passed away since the occurrence and true character of those organic remains were ascertained by Mr. Smith of Jordanhill. Cruising with his yacht among the kyles and lochs of his own Firth of Clyde, he had been collecting materials from the raised beaches of the west to show that the land had undergone a comparatively recent elevation. One day, in company with a friend, he chanced to walk along the beach of a little bay in the Kyles of Bute. His attention was directed to a number of shells lying among the shingle, but different from any which his companion or himself had ever dredged in the adjoining sea. On closer inspection it was found that the shells had been washed out of a bed of clay, where they existed by hundreds, and that their association on the beach with the recent shells thrown up by the tides was merely an accident. What then constituted the difference between the shells of the clay bed and those living in the neighbouring kyles and firths? It was at first supposed that some of them were of new species. But by degrees it was ascertained that they all belonged to species yet living, that some of them were still natives of the seas of Scotland, but that others were now to be found only in the seas of Norway and the Arctic circle. Having once fully grasped this fact, Mr. Smith was not slow to perceive its significance. He traced the clay-bed along many parts of the west coast, and, in order the better to compare its contents with the shells still inhabiting the British Seas, he instituted a careful dredging of the basin of the Firth of Clyde. A more charming employment can hardly be conceived. In the midst of some of the finest scenery on the west of Scotland, within easy reach of all the comforts of home, and yet among scenes almost as lone and retired as the wildest Highland tarn, his self-imposed task was to bring up to the light of day the denizens of these quiet sea-lochs and bays, to explore the deeps and shallows, sunken reefs, shoals, and abysses, and thus, while his vessel perhaps lay asleep on the face of a summer sea, to walk, as it were, in fancy along the sea-floor many fathoms underneath; and to pick up there, from its nestling place among tangle or coral sand, many a tiny shell which had never before been known to live around the coasts of Britain. The result of his explorations, and of those subsequently carried on by Forbes, Macandrew, Jeffreys, and others, went to show that the assemblage of shells in the clay had a strongly northern character, that among them were some which are now rare in British seas, though common in the far north, and that fourteen or fifteen species no longer live around our shores, but are confined to the boreal and arctic regions. The value of these researches, made fully five and twenty years ago, was not thoroughly perceived at the time, but they were eventually found to lend a powerful support to the attempts of geologists to account for some of the supercial phenomena of the country by the agency of ice.

Neither the boulder-clays nor the brick-clays form any prominent features in the landscape. They are spread out either in flat sheets covered with fields or in smooth rolling ridges. It is only where the underlying rock rises through this wide mantle that the monotony and tameness of the surface is interrupted. Above the boulder-clay, however, there occur two deposits, which play a minor but not unimpressive part in many a piece of Lowland scenery. These are mounds and ridges of sand and gravel, and scattered boulders or erratic blocks.

Among the formations which overlie the till, and belong to the closing ages of the glacial period, must be placed those long rampart-like ridges of gravel and sand, known as *Kames* in Scotland, *eskers* in Ireland, and *ösar* in Scandinavia. Notwithstanding all that has been said and written about them, they are as complete a mystery as ever to the geologists of this country. They look not unlike the earthen mounds of some antique fortification, only that they are greatly loftier, longer, and not less perfect than any such fortification which has survived in this country. They rise up sharply and boldly, sometimes from the side of a hill, sometimes along a wide moor, and sometimes across a valley. They do not appear to occur in Scotland except in the neighbourhood of hills and rising ground. They may be traced at all levels—from less than a hundred feet above the sea up to at least a thousand feet. They consist of sand or of gravel, or of both, varying in texture to the coarsest shingle, without fossils, save on the Aberdeenshire coast, where they have yielded a few sea-shells. These ridges have been a fruitful source of wonder and legend to the people. It was a quaint and beautiful superstition that peopled such verdurous hillocks or *tomans* with shadowy forms, like diminutive mortals, clad in green silk, or in russet grey, whose unearthly music came sounding out faintly and softly from underneath the sod. The mounds rose so conspicuously from the ground, and, whether in summer heat or winter frost, wore ever an aspect so smooth and green, where all around was rough with dark moss-hags and moor, that they seemed to have been raised by no natural power, but to be in very truth the work of fairy hands, designed at once to mark and guard the entrance to the fairy world below. The hapless wight who, lured by their soft verdure, stretched himself to sleep on their slopes, sank gently into their depths, and after a seven years' servitude in fairy-land awoke again on the self-same spot. Like young Tamlane,

> "The Queen of Faeries keppit him
> In yon green hill to dwell."

According to a tradition in Roxburghshire, the Kames are the different strands of a rope, which a troublesome elfin was commanded to weave out of sand. The strands were all prepared, but when the imp tried to entwine them, each gave way, and hence the broken parts of the Kames have remained to this day. Michael seems to have had no small amount of work in altering the surface of the country. There is a deep gash through a sandy ridge—part of the Kame series—at the south end of the Pentland Hills, and not far off stands a green conical sand-hill. The wizard is said to have dug the trench and piled up the hill in the course of a single night. It was he, too, that

> "Cleft the Eildon Hills in three,
> And bridled the Tweed with a curb of stone."

Throughout the south of Scotland the more obtrusive minor features of the scenery are often traced up to the agency of Michael Scott and his band of witches and warlocks. Fanciful and sometimes grotesque as these legends are, they are yet interesting, inasmuch as they indicate the prominence of the phenomena, and the difficulty of accounting for them by any of the common operations of nature.

That the Kames are connected in some way with the action of ice is shown by the fact that they disappear as we advance southward from Scotland through the northern counties of England, and by the occurrence of occasional striated stones in them, and of large boulders lying upon them. But that they are not the ordinary moraines of glaciers, as some geologists have imagined, seems to be conclusively indicated by the absence of angular rubbish, by the well-worn water-rolled character of the stones, and by the stratification which is almost everywhere visible in them, when a sufficiently large section is exposed. Moving water, therefore, must also have been concerned in their production. He will be a lucky observer who succeeds in harmonizing the difficulties, and presenting a satisfactory explanation of these remarkable ridges.

Connected with the Kames, and perhaps nearly as old, is a series of tarns and of former lake-basins

FIG. 23.—SECTION OF SAND AND GRAVEL RIDGES (KAMES) AT CARSTAIRS, LANARKSHIRE. (The dark portions mark little basins of peat occupying the site of former tarns.)

now filled with peat. I do not know a district where these features play so conspicuous a part in the scenery as in the eastern parishes of the county of Lanark. Behind the little village of Carstairs, for instance, the ridges of sand and gravel run one after another from south-west to north-east, somewhat like the larger mounds of a tract of sand-hills by the sea. Instead of following straight parallel lines, however, they are singularly tortuous, so that they often come together, and in this way form loops, which enclose basins of water or of peat. One such hollow in particular is so circular, and shelves so steeply into the pool which fills its bottom, that it at once suggests the crater of a volcano like one of those in the Eifel. As the Kames there stretch across the mouth of a broad valley, they must at one time have dammed back the drainage so as to form a lake. Since then they have been cut through by the Mouse Water, and the lake has thus been drained. But its site is still visible in the wide moss-hags and bogs of the Carnwath Moor, and in at least one place a shrunk remnant of the water, with the peat creeping into it, may even yet be seen. The gradual inroads of the peat upon the smaller ponds and lochans is also well exhibited. Standing on the crest of one of the higher ridges, the observer can at once understand how, after the formation of these mysterious mounds, there must have been dozens of little tarns or pools lying in dimples and basins among the Kames. But he can see only three or four which have not been converted into peat-bogs.

The last illustration which I shall give of the influence of the various agencies of the glacial period upon the scenery of the Central Valley of Scotland is to be found in what are known to geologists as *Travelled* or *Erratic Blocks*. Scattered over the island from sea to sea are numberless boulders of all sizes, up to masses of many tons in weight. Thus

> " A huge stone is sometimes seen to lie
>   Couched on the bald top of an eminence;
>   Wonder to all who do the same espy,
>   By what means it could thither come and whence,

> So that it seems a thing indued with sense,
> Like a sea beast crawled forth, that on a shelf
> Of rock or sand reposeth, there to sun itself." [1]

Unhappily, the progress of modern agriculture is inimical to the preservation of these stones, and they have as a consequence disappeared from the more cultivated districts. But in many a mossy tract, especially round the flanks of the main hill ranges, they may still be counted by the score. To the north of the higher hills of Carrick, indeed, the grey granite boulders lie strewn on the ground by hundreds, and when seen from a little distance they look like flocks of sheep. So conspicuous are the erratic blocks of the Lowlands as to have long attracted the notice of the peasantry, and so strange sometimes are their positions, and so markedly do they often differ in composition from the general character of the surrounding rocks, that, like the Kames, they have been from the earliest times a theme of endless wonder. Many a wild legend and grotesque tale of goblins, witches, and elfins has had its source among the grey boulders of a bare moor.[2]

---

[1] Wordsworth's *Poems of the Imagination*, xxii.

[2] In wandering over the south of Scotland I have met with some curious traditions of this kind. The following was told me on the spot by an intelligent native of the village of Carnwath. Before farming operations were there carried to the extent to which they have now arrived, large boulders, now mostly removed, were scattered so abundantly over the mossy tract between the river Clyde and the Yelping Craig, about two miles to the east, that one place was known familiarly as

"Giant's Stone," "Giant's Grave," "Auld Wives' Lift," "Witches' Stepping Stanes," "Warlock's Burdens," "Hell Stanes," and similar epithets, are common all over the Lowland counties, and mark where to the people of an older time the singularity of these blocks proved them to be the handiwork, not of any mere natural agent, but of the active and sometimes malevolent spirits of another world. Nor need this popular belief be in any measure a matter of surprise. For truly, even to a geological eye, which has been looking at the same phenomenon for years, each fresh repetition of it hardly diminishes the interest, nay, almost the wonder, with which it is beheld.

as "Hell Stanes Gate" [road], and another "Hell Stanes Loan." The traditional story runs that the stones were brought by supernatural agency from the Yelping Craigs. Michael Scott and the Devil, it appears, had entered into a compact with a band of witches to dam back the Clyde. It was one of the conditions of the agreement that the name of the Supreme Being should never on any account be mentioned. All went well for a while, some of the stronger spirits having brought their burden of boulders to within a few yards from the river, when one of the younger members of the company, staggering under the weight of a huge block of greenstone, exclaimed, "Oh, Lord! but I'm tired." Instantly every boulder tumbled to the ground, nor could witch, warlock, or devil move a single stone one yard further. And there the blocks lay for many a long century, until the rapacious farmers quarried them away for dykes and road-metal.

Another explanation of a somewhat different kind was given by a stonemason among the Carrick Hills, who, on being asked how he imagined that the hundreds of granite boulders in that district came to lie where they do, took a little time to reply, and at last gravely remarked that he "fancied when the Almichty flang the warld out, He maun hae putten thae stanes upon her to keep her steady."

We have rudely dispossessed the old warlocks and brownies; and yet, though we can now trace, it may be, the source from which the stones were brought, and the manner in which they were borne to their present sites, their history still reads like a very fairy tale. There they lie crusted with mosses and lichens, and with tufts of heather, and harebell, and fern nestling in their rifts, while all around perhaps is bare bleak moorland. How came they there? They have not tumbled from any cliff, for we may see them rising boldly above the soil, when not another vestige of naked rock appears within sight. They have not been transported by rivers, for they are often seen perched on the summits of the hills, high above all the streams, and even out of hearing of their sound. They cannot have been washed up by floods and oceanic convulsions, for not only are they in many cases of enormous size, but they consist of rock which is frequently foreign to the district, and may not be found nearer than fifty or sixty miles, beyond successive ranges of hills and valleys. What force, then, could carry these huge masses to such great distances across wide and deep valleys, and lines of high hills? Again we must answer, Ice.

Scattered over the chains of the Sidlaw and Ochil Hills lie large masses of gneiss and schist that have come across from the opposite ranges of the Grampians. Boulders of the same kind are likewise found both on the plains and hills to the south of the Forth. So from the granite mountains of Carrick and Galloway, millions of boulders have been strewn over the heights and hollows to the north-west and north, up to, and perhaps beyond, the town of Ayr. It can hardly be doubted that these large masses of stone have been carried on floating ice across the sea, and dropped on the surface of the submerged land. They must thus be assigned to that part of the long glacial period when a large area of the country had sunk beneath the waves, and the cold still continued so severe, as to keep snow-fields and glaciers among the mountains, and even, perhaps, to freeze in winter the water of the ocean. It was then that the Lowland Valley of Scotland, and the less elevated portions of the rest of the kingdom, were covered with a sea, across which bergs and ice-rafts were ever drifting to and fro, grating along the bottom, and carrying with them the *débris* of many a Highland glen and shore, to drop it at random over the drowned land.

When the great Ice Age came to a close, and this country had once more risen above the sea, the lower grounds, especially throughout the Midland Valley, must have worn a very different aspect from that which had distinguished them before either glacier or iceberg had begun to modify their surface. Their old pre-glacial roughness had been, to a large extent, planed down, and new hollows had doubtless been scooped out by the ice; but a thick mantle of clay and sand had been spread over the ground, leaving only the harder and higher masses of rock to rise above the general monotonous undulations. We have seen how in the case of the Highlands and of the Southern Uplands the atmospheric forces of waste have been busy since these tracts re-emerged, how they have trenched the hill-sides with water-courses, and dug ravines in the valleys, and how they are still everywhere at work in destroying the smoothed outline that was left by the ice. The same changes are likewise going on in the Lowlands. The wide covering of drift which there overlies the country has been greatly worn. It is furrowed in all directions by runnels and brooks; the larger streams have cut ravines through it down to the solid rock below, nay, in numberless cases they have even hollowed out deep gorges in the rock itself. And if we turn to the hills that rise out of the wide plain of drift, and lift their bare rocks to the sky, proofs of the same waste meet us on every side. The hummocks of greenstone are split and broken, the crags of basalt are rent and splintered; the base of the cliffs and the sides of the hills are cumbered with the ruin that marks how quietly, yet how well, the rains, springs, and frosts of centuries have done their work. Slowly the impress of the ice is fading away, and though thousands of years may pass before all trace of its action is obliterated, the time must nevertheless come when the present surface of the land shall have disappeared as completely as the sand ripple of last night's tide was washed away by the tide of this morning.

CHAPTER XII.

CHANGES IN THE SCENERY OF SCOTLAND SINCE THE ADVENT OF MAN.

AMONG the more marked changes which have influenced the scenery of the country since the close of the glacial period, let me refer, in conclusion, to two that are intimately linked with the human history of the island—the *Raised Beaches* and *Peat Mosses*.

RAISED BEACHES.

If, after its submergence, the country rose again slowly, with long intervals of rest, each of these pauses would give the sea an opportunity of cutting a notch, or horizontal terrace, along the margin of the land. A succession of such terraces, or "raised beaches," might thus be traceable at different elevations above the present sea-level, becoming generally fainter according to their height and their consequent antiquity. The thickness of the coating of snow and ice which still enveloped the rising land, seems, however, to have either prevented these indentations from being distinctly made, or to have in great measure effaced them. For, save in a few exceptional cases, as near St. Andrew's, where well-marked terraces are seen up to a height of 290 and even 350 feet, it is only the lower and more recent examples which form conspicuous features in the landscape. Round the west coast a terrace, having a height of about forty feet above high water-mark, winds as a green platform along the dark rocky coast of Argyleshire and Inverness. It probably dates from the later part of the glacial period, for Arctic shells have been found in it. It is succeeded by a lower and later terrace, long ago described by Mr. Smith, of Jordanhill, and Mr. Maclaren, which runs at a height of about twenty or twenty-five feet above high water. This, on the whole, is the most marked of all the raised beaches, for it is of comparatively recent date. As it has yielded in several places works of human fabrication, it must be of later date than the time when man became an inhabitant of this island. From the nature of these remains, and other evidence, I have been led to infer that its formation has taken place, either in whole or in part, since the first century of our era.[1]

Round the greater part of the sea-margin of Scot-

---

[1] *Edin. New Phil. Journ.* for 1861, and *Quart. Journ. Geol. Soc.* for 1862. My inference, however, has been disputed by Mr. A. Bryson (*Edin. New Phil. Journ.* for 1862,) and Mr. W. Carruthers (*Quart. Journ. Geol. Soc.* 1862).

land this terrace runs as a flat selvage of sandy, gravelly, or clayey ground, varying in breadth from six or seven miles to not more than a few feet. It rises from twenty to thirty feet above high watermark, and is composed of horizontal layers of sand, gravel, or clay, often full of littoral shells, the whole having been laid down by the sea. Along the inner margin of the terrace the ground usually rises as a line of shelving bank or precipitous cliff, just as a shelving bank or steep cliff shoots upward from the sea. Moreover, the inland cliff that bounds so many portions of the terrace is not unfrequently scarped into clefts and creeks, and perforated with long dim caverns. The resemblance to a sea-cliff goes still further, for the terrace itself is often dotted with prominent crags and worn pillars of rock, like the tangle-covered skerries and sea-stacks that roughen many a wild beach, open to the full swell of the North Sea or the Atlantic. These inland rocks, indeed, whether on the terrace or rising steeply from its inland edge, are feathered over with ferns and ivy and trailing briers; they are tinted with mosses and lichens, and gay with many a bud and blossom, luxuriant bunches of hart's tongue hang from the roofs of the caves, and swallows build their nests in the crannies of the cliff. But divest the rocks of all this tapestry of verdure, strip the terrace of its mantle of gardens and fields, its highways and hedgerows its villas and hamlets, busy sea-port towns and watering-places, and you then lay bare a sandy flat that ends at the foot of a gentle slope, or at the base of a line of bleached and wasted rocks. You, in fact, reconstruct an old coast line, and there can be no more doubt that the sea once rolled over that terrace, and broke against that cliff, than that the waves are breaking over the beach to-day. Instead of the level corn-fields and orchards of the terrace, imagine a tract of sand or mud; for the mosses and lichens, ferns and flowers, substitute a shaggy covering of sea-weed; in place of swallows, martins, and rock-pigeons, people the rocks with gulls, and auks, and cormorants; let the tides come eddying across the terrace among the rocks and the cliff: and you thus restore that old coast-line to the condition in which it existed at a comparatively recent geological period. If you could gently depress the land for some twenty or thirty feet you would actually bring back the old outline of the Scottish shores ere the last upheaval had begun.

Evidence of the truth of these remarks must be familiar to every one who has visited almost any part of the coast-line of Scotland. The old or upraised beach runs as a terrace along the margin of the Firth of Forth; it forms the broad carses of Falkirk and of Gowrie; it is visible in sheltered bays along the exposed coasts of Forfar, Perth, Kincardine, Aberdeen, and westwards along the Moray Firth. On the Atlantic side of the island, its low green platform borders both sides of the Firth of Clyde, fringes the islands, runs up the river beyond Glasgow, and winds southwards along the coast of Ayrshire and Wigtown into the Irish Channel.

This great terrace cannot be accounted for in any other way than by admitting that it was formed by the action of the sea, and that since its formation there has been a rise of the land to a height of from twenty to thirty feet above the level which it previously occupied. This upheaval was of course brought about by the operation of those igneous forces that are lodged within the earth, but whose origin and mode of working still remain such a mystery. It probably went on slowly, the land rising inch by inch, and foot by foot, just as the coast of Sweden is rising at the present time. At last the elevation ceased, and the old beach, now high above the tides, became a level platform coated with turf. Few geological changes have been more directly serviceable to man. He has found this upraised sea-margin an inestimable site for his maritime towns and villages; and along many a coast-line, where, before the elevation, the ground shelved down in a cliff or steep bank to the sea, he now finds ready to his hand a firm level terrace, bounded by the former cliff on the one side, and by the present sea-beach on the other.

Leith, Burntisland, Dundee, Arbroath, Cromarty, Rothsay, Greenock, Ardrossan, Ayr, and many other towns on the coast stand, either wholly or in part, on this terrace. Indeed it is in no small degree owing to the facilities afforded by the terrace that the banks of the river and Firth of Clyde are so thickly fringed with towns, villages, and watering places. At Glasgow, which is partly built on the same platform, some interesting relics of the early history of these geological and historical features have been found. From the silt and sand of which the terrace there consists, no fewer than eighteen canoes have at different times been obtained, some of them from under the very streets and houses. Apart from questions of science, it is not uninteresting to mark at how early an epoch the advantages of the Clyde, as a maritime station, were recognised. The number of canoes seems to show that the river was much frequented, although no record remains to indicate what may have been the traffic in which they were engaged.

I have already alluded to the singular contrast between the present aspect of the Clyde and its appearance during the bleak glacial period. Another, not less striking in its features, and bearing a closer human interest withal, is suggested by these relics of the early races. To-day all is bustle and business. Ships from the remotest corners of the earth come hither with their merchandise. Vast warehouses and stores are ranged row upon row along the margin of the river, and in these are piled the productions of every clime. Streets, noisy with the rattle of wheels, and the tread of horses, and the hum of men, stretch away, to the right hand and the left, as far as the eye can reach. The air is heavy with the smoke belched out from thousands of chimneys. And so, day after day, the same endless din goes on; every year adding to it, as the streets and squares creep outward, and the tide of human life keeps constantly flowing. But how different the scene when our hatchet-wielding forefathers navigated these waters! Down in the earth, beneath those very warehouses and streets, lies the bed of the old river, with the remains of the canoes that floated on its surface—silent witnesses of the changes that have been effected, not less on the land than on its inhabitants. We can picture that dim, long-forgotten time, when the sea rose at least five-and-twenty feet higher in the valley than it does now, and covered with a broad sheet of water the site of the lower parts of the present city of Glasgow. We see the skirts of the dark Caledonian forest sweeping away to the north, among the mists and shadows of the distant hills. The lower grounds are brown with peat bogs and long, dreary flats of stunted bent, on which there grows here and there a hazel or an alder-bush, or, perchance, a solitary fir, beneath whose branches a herd of wild cattle browse on the scanty herbage. Yonder, far to the right, a few red deer are pacing slowly up the valley, as the heron, with hoarse outcry and lumbering flight, takes wing, and a canoe, manned by a swarthy savage, with bow across his shoulders, pushes out from the shore. The smoke that curls from the brake in front shows where his comrades are busy before their huts hollowing out the stem of a huge oak, that fell on the neighbouring slope when the last storm swept across from the Atlantic. And there stretches the broad river—its surface never disturbed save by the winds of heaven, or the plunge of the water-fowl, and the paddles of the canoes—its clear current never darkened except when the rain clouds have gathered far away on the Southern hills, and the spate comes roaring down the glens and waterfalls, and hurries away red and rapid to lose itself in the sea. Such was the landscape when our ancestors first looked upon it. How came it to undergo so total a change? It is not merely that man himself has advanced, that he has uprooted the old forests, extirpated the wild cattle, driven away the red deer to the fastnesses of the mountains, drained the peat-bogs, covered the country with corn-fields and villages, and built along the margin of the river a great city. True, he has done all this, and has undoubtedly been the chief agent in the general change. But nature, too, has helped him. Those vast forces that are lodged beneath the crust of the earth have slowly upheaved the land, and have converted a large part of the bottom of the old estuary into good, dry ground, covered with the richest soil, and fitted in no common degree for the growth of streets. And hence, where his forefathers floated their rude boats he builds his warehouses, and on tracts that were ever wet with the ooze of river and sea, and bore few other inhabitants than the cockle and mussel, he now plants his country villas and lays out his pleasure-grounds.

DISAPPEARANCE OF THE ANCIENT FORESTS; GROWTH OF PEAT MOSSES.

The disappearance of the ancient woods deserves more than a mere passing allusion, for it has materially influenced the present scenery of the country, and it has a still further interest from the close way in which it is linked with human history. Duly to appreciate the nature and extent of the change which is traceable to this cause, it is necessary to bear in mind the magnitude of the forests which, when man first set foot in Scotland, swept in long, withdrawing glades across its surface,—of the wide black mosses and moors,—of the innumerable lakes and fens, dense and stagnant indeed on the lower grounds, but which, in the uplands, were the sources whence streamlets and rivers descended through glen, and valley, and dim woodland, into the encircling sea. Beasts of the chase, and among them some that have been for centuries extinct here, abounded in these ancient forests; birds of many kinds haunted the woods and waters; fish swarmed in lake, river, and sea. Among such primeval landscapes did our aboriginal forefathers excavate their rude earthen dwellings and build their weems of stone; from the stately oaks they hollowed out canoes, which they launched upon the lakes and firths; and through the thick glades of the forest they chased the wild-boar, the *urus*, the bear, the wolf, and the red deer. The traces of these old scenes are still in part preserved to us. From the lakes and peat-mosses are sometimes exhumed the canoes, stone celts, bronze vessels, and ornaments of the early races, along with the trunks of oak and pine that formed the ancient forest, and bones of the animals that roamed through its shades. It is from such records that we know both what used to be the aspect of the country and how it has come to be so wholly changed.

It is a common opinion that the peat-mosses of Scotland are of a comparatively modern date—not older, indeed, than the Roman invasion, because "all the coins, axes, arms and other utensils found in them are Roman." But these relics are better understood now than they were some years ago; and though in some cases their Roman date is beyond doubt, they are admitted to belong generally to the earlier time, known to the antiquary as the Bronze Period. Their evidence, therefore, cannot prove more than that the mosses in which they have been found may be later than the time when the natives of this country fashioned their implements of bronze:—*may* be later; for the occurrence of the antiquities in the peat is of itself no proof that the peat is not actually very much older than they. They may in fact have been dropped on the moss when it was in a soft, boggy condition, and so have sunk to some depth beneath the surface. It would require not a little careful observation to show conclusively that the portion of the peat lying above such remains was really formed after they were left there by their human owners. If, however, the remains occur not in the substance of the moss, but below it, on what was once a soil, or a lake-bottom, and if they are of such a kind, or in such a position, as to show or to make it probable that they were left exactly in the place where they still lie, the inference may be drawn that they are of older date than the peat which overlies them. Tried by such a rigid test as this, comparatively few of the Scottish peat-mosses can at present be proved to be later than the Roman invasion. There is ground for believing that a good many, however, have probably been formed since that time; others, though later than the first coming of man into the country, must be far older than our era; while some may even go back into the glacial period. It is in the depths of this last-named series, and in the beds of clay and marl underneath which mark the bottoms of former lakes, that we may look with most hope for the discovery of the remains of the animals that first inhabited the country, after it rose out of the glacial sea. No opportunity should be lost of watching the cutting and draining of such peat-mosses. It is there that we may expect to meet with the great extinct Irish elk, and with the progenitors of our present races of cattle; and there, too, we may chance to light on the mammoth, rhinoceros, reindeer, musk-ox, bear, and other animals which preceded or may have been contemporary with the earliest human population of these wilds.

Peat can only be formed in lakes, or on wet, marshy ground. In the former case the water is gradually displaced by the growth of marsh-plants, creeping steadily from the margin to the centre, until a surface has been formed over the site of the old lake. This process may be seen going on in many parts of the country; and so rapidly does it sometimes advance, that the sheet of water becomes almost visibly smaller every year, while the encircling morass gains in proportion. Such seems to have been the origin of not a few of our peat-mosses, and especially of the older ones. The pools and lakes formed by the unequal accumulation of the deposits of the glacial period have in many cases passed into basins of peat; and the cotter now cuts his fuel where of old the wild duck made its nest. Some of these peat deposits, for ought we can tell, may be almost as old as the time when the basins in which they lie were first filled with fresh water. There are proofs, however, that some of the basins were still lakes when man was living in the island, for canoes have been found beneath several feet or yards of peat, and lying on fine sand and gravel, evidently the bottom of the old lake.

Peat-mosses not only mark the site of lochs and tarns; they cover the ruins of ancient woodlands. That the mosses which stretch over so many wide moors in Scotland have sprung up after the destruction of forests which once grew there, is shown by the numerous trunks and branches of trees which are so constantly found among the lower parts of the peat. A thick grove of oaks would not be likely to spring up on the surface of a quaking bog. On the contrary, it was owing to the destruction of the forests that the bog arose. The prostrated trunks would of course intercept any of the little runnels that might have been wont to trickle through the woods, and thus stagnant swamps would be formed, in which water-mosses would readily take root; and there would grow up in this way a true peat-moss. When we consider how large an extent of surface was covered with wood in the early times of history, and how much of what must then have been woodland is now morass and bog, slowly being reclaimed by the farmer, there is ground for the inference, as has just been said, that a good deal of this kind of peat must be later than the days of the Romans.

There are several ways in which a forest may be destroyed and turned into a peat-moss. The growth of a thick mass of wood for many successive centuries on the same spot, tends to impoverish the soil, and, in the natural course of events, the trees must decay and give way to other races of plants, which will draw nourishment from the mouldering trunks. And thus, on tracts which at one period bore a dense array of wood, there might spring up in later ages long brown morasses and peat-bogs. Again, a fierce hurricane sweeping across the country may prostrate the trees over wide areas; and the fallen trunks and rotting leaves, by collecting moisture and facilitating the growth of marshy vegetation, may in like manner give rise to a peat-moss. Or the weight of snow in a severe winter may be so great as to break the branches, and even drag down the trees upon each other: *or the forest may be destroyed by fire.*

> " Nec jam sustineant onus
> Sylvæ laborantes, geluque
> Flumina constiterint acuto."[1]

Or, lastly, man, armed with axe and hatchet, may come and fell oak and beech and pine, taking, it may be, little or none of the wood away, but leaving it there to rot, and to gather around and over it a mantle of peat-forming plants. Peat-mosses appear to have arisen in each of these ways in Scotland. In the Forest of Mar, Aberdeenshire, large trunks of Scotch fir, which fell from age and decay, were soon immured in peat, formed partly from the decay of their perishing leaves and branches, and partly from the growth of *sphagnum* and other marsh-plants. About the middle of the seventeenth century, a storm swept down from the mountains of Loch Broom, in Ross-shire, and levelled a forest so completely, that in less than fifty years thereafter peat was dug from the same spot. In 1756 a similar fate was experienced by the whole Wood of Drumlanrig, in Dumfriesshire. And other cases are known where, at the bottom of the moss, lie the remains of old forests, with their trees prostrated all in one direction, showing the point from which came the storm that hurled them to the ground.[2]

The Moss of Kincardine, in the upper part of the

[1] Hor. *Carm*. I. ix. 2.
[2] Rennie's *Essays on Peat*, pp. 30, 65.

valley of the Forth, owes its existence to the fact that the thick oak forest which once covered these grounds has been felled by man. Below the moss the stumps and trunks of large trees were found crowded as thickly upon the clay as they could be supposed to have grown there. The roots were still fixed in the clay, as when the trees were in life, and the stems had been cut down at a height of about two feet and a half from the ground. Marks of a narrow axe were sometimes traced on the lower ends of the logs, completing the proof that the wood had been cleared by human agency.[1] Here we see how a district of fair woodland—the home, doubtless, of many a stag and hind, and the nesting place of many a cushat dove and blackbird—has been turned by man into a waste of barren morass and mire—a place of shaking bog and stunted heath, where he cannot build his dwelling nor plant his crops, and from which he can extract nothing save fuel for his hearth. Such has been the condition of these districts for many a long century; and it is only within the last two or three generations that an exertion has been made, with much labour and cost, to strip off the covering of peat, and restore again to the light of day that old soil which nourished the early oak forest.

It is curious to mark on many a peat-moss, particu-

---

[1] Tait, *On the Mosses of Kincardine and Flanders.* Trans. Roy. Soc. Edin. iii.

larly on those along the crests of hills, how the same universal system of decay, which we have now traced among the solid rocks and surface soils over the whole country, is at work in eating away the black peat, and washing its waste into the valleys. It is to this cause that the singularly rugged surface, known in the south of Scotland as *moss-hags*, is due. The peat is no longer growing, or at least grows so feebly that it cannot repair the damage done by rains and frosts. Hence deep gutters and pools are dug out of the crumbling mass—black, soft, and treacherous, only passable in dry weather, and even then often like the march of the Salian priests, "*cum tripudiis sollemnique saltatu.*" Why the peat-mosses should thus die out is a question deserving of inquiry. Although there may be some more general cause, the draining of the country in agricultural operations has doubtless had a material effect upon the mosses. Throughout the uplands of the southern counties, the black cappings of peat which cover so many of the flat hill-tops, and extend down their sides, may now be seen to be shrinking up again towards the top. They have a ragged fringe, some parts running in long tongues down the slope, or in straggling isolated patches. These features are well seen from the high grounds above Loch Skene. The long, bare, flattened ridges have each their rough scalps of peat, of which the black, broken edges hang down the slopes of brown heath and bent, while far below are the green valleys, with their clear winding streams, and their scattered shepherds' hamlets.

INFLUENCE OF MAN UPON THE SCENERY OF THE COUNTRY.

It would lead to too wide a discussion to enter fully here upon the influence of man in bringing the scenery of the country to its present condition. He has uprooted the old forests, drained many of the mosses, and extirpated or thinned many of the wild animals of ancient Caledonia. In place of the woods and bogs, he has planted fields and gardens, and built villages and towns; instead of wild beasts of the chase, he has covered the hills and valleys with flocks of sheep and herds of cattle. The cutting down of the forests and the draining of the mosses has doubtless tended to reduce the rainfall, and generally to lessen the moisture of the atmosphere and improve the climate. Sunlight has been let in upon the waste places of the land, and the latent fertility of the soil has been called forth; so that over the same regions which, in Roman times, were so dark and inhospitable, so steeped in dank mists and vapours, and so infested with beasts of prey, there now stretch the rich champagne of the Lothians, the cultivated plains of Forfar, Perth and Stirling, of Lanark and Ayr, and the mingling fields and gardens and woodland that fill all the fair valley of the Tweed, from the grey Muirfoots and Lammermuirs far up into the heart of the Cheviots.

In effecting these revolutions, man has introduced an element of change which has extended through both animate and inanimate nature. He has ameliorated the climate, and by so doing has affected the agencies of waste that are wearing down the surface of the land. The rivers are now, probably, a good deal less in size than they were even in the days of the Romans, and there may be fewer runnels and streamlets. The old mosses acted as vast sponges, collecting the rain that fell upon them or soaked into them from the neighbouring slopes, and feeding with a constant supply the brown peaty rivulets that carried their surplus waters to the lower grounds. The evaporation from these wide swampy flats could not but be extensive, and the rain-fall was thus, in all likelihood, proportionately great. But the clearing away of the forests and of the peat-mosses has removed one chief source alike of the rivulets and of the rain. The amount of denudation by the combined influence of rain and streams ought accordingly to be less, on the whole, than it was eighteen hundred years ago. At the same time, it should be borne in mind that the extent to which draining has now been carried all over the country has had the effect of allowing the rain to run off more easily into the rivers. Hence the latter swell and fall again more rapidly than they used to do. Floods or "spates," though the rainfall may be the same or less, have thus a tendency to be more sudden and violent than formerly, and hence, in the increased amount of erosion performed by the river in flood, there may, perhaps, be an equivalent for the diminution of the stream in its ordinary state.

Among the plants and animals of the country, too, traces of the influence of man's interference are everywhere apparent. He has altered the character of the vegetation over wide districts, driving away plants of one kind, such as the heaths, to put in their stead those of another type, like the cereals. The gradual change of climate must also have affected the distribution of the vegetation of the country: some herbs grow now more abundantly than they did before; or they may now be able to flourish at a higher level than of old. Others, to which the change has been unfavourable, may have been greatly thinned in numbers, and even extirpated altogether. In like manner the coming of man has worked mighty transformations in the animal world. Over and above the extirpation of the beasts of the forest, and the introduction of foreign forms into the country, he has waged incessant war against those which he considers injurious to his interest. He has thus altered the natural proportion of the different species to each other, and introduced a new element into the universal "struggle for existence." No species, whether of plants or of animals, can notably increase or diminish in number without, of course, thereby exerting an influence upon its neighbours. And here a boundless field of inquiry opens out to us. Man's advent has not been a mere solitary fact, nor have the alterations which he has effected been confined merely to the relations that subsist between himself and nature. He has set in motion a series of changes which have reacted on each other 'in countless circles, both through the organic and the inorganic world. Nor are they confined to the past; they still go on; and, as years roll away, they must produce new modifications and reactions, the stream of change ever widening, carrying with it man himself, from whom it took its rise, and who is yet in no small degree involved in the very revolutions which he originates.

CHAPTER XIII.

RECAPITULATION AND CONCLUSION.

IN this final chapter let me present the reader with a brief summary of what has been said in the chapters that have gone before. At the outset I tried to show that the common and popular notion which assigns the present inequalities on the earth's surface indiscriminately to the results of early earthquake and upheaval is untenable; that the only principles on which we can advance with confidence in seeking to decipher the history of our hills and valleys are those laid down by Hutton and his illustrator, Playfair; and that by studying the different agencies at work in altering the face of the globe at present, we take the only way open to us of investigating the progress of change in the geological past. In this course we are led to perceive that, although there have been many upheavals, depressions and fractures of the earth's crust, nevertheless the present inequalities of the land are probably due in the main to the unequal waste of the rocks by rains, springs, streams, ice, and the sea; and that, as Hutton long ago complained, it is inability or unwillingness to grant the enormous periods of time demanded by the working of these slow but constant forces, which has long stood in the way of the general acceptance of this part of the Huttonian Theory.

When, in accordance with these principles, we begin to inquire into the origin of the present scenery of the country, we are soon taught that each hill and valley, each mountain and glen, has a twofold history. There is first the story of the formation of its component rocks, whether these have been laid down layer after layer as sand, gravel, or mud upon the bottom of a former sea, or piled up as shingle along an ancient beach, or drifted as the finest ooze over the bed of a lake; whether formed of the decay of extinct forests, or from the gathered fragments of corals and shells; whether rolled along in the form of liquid lava, or thrown up in showers of volcanic dust and ashes. And after we have tried to trace out the succession of events imperfectly chronicled in the rocks, and have learnt, in so doing, how little we know, and how utterly beyond human realization is the vastness of the antiquity thus recorded, there still remains the story of those after changes, whereby the various rocks that were piled over each other came to be upheaved and carved into the present framework of the country. Between the time when the rocks were formed and that in which they were raised into the land on which our hills and valleys have been moulded, long millions of years must in many cases have passed away, during which metamorphism and other underground processes were at work; for when these rocks appeared in the light of day they were often vastly different from their original condition. Sand, silt, and mud had been changed into schist, slate, gneiss, and granite; and this not in mere local patches, but, as in the Highlands, over an area many thousands of square miles in extent.

The hills and valleys of Scotland, we have seen, are not all of one age. They differ greatly also in geological structure, with a corresponding variety of scenery. As a convenient subdivision they were grouped into three districts,—the Highlands, the Southern Uplands, and the Midland Valley. In taking leave of them, however, for the present, let us regard them finally as a whole, and picture briefly the changes by which their rocks, whether formed in lake, river, or sea, have come at last to wear their present outlines on the surface of the land. We watch them raised by subterranean movements within reach of the waves, and there for long ages

"Swilled with the wild and wasteful ocean,"

until hundreds and thousands of feet of solid rock had been worn away. During possibly many risings and sinkings of the land this marine denudation went on, and by degrees the waves succeeded in levelling the country into broad undulating table-lands. It was out of such sea-worn platforms that the Scottish mountains and valleys appear to have been carved. We mark how simply the present grouping of the valleys may have arisen. Rain falling on the land that was rising above the sea-level, found its way from the centre by devious paths outwards and downwards to the shore. These paths, once chosen, would ever be deepened and widened as century after century rolled away. The wide table-lands, like a sandy beach on the recession of the tide, were thus slowly hollowed out on every side by little runnels, that gathered into brooks, thence into larger streams, and lastly into broad rivers that swept the drainage out to sea. Year by year the process of excavation went on, every shower of rain, every spring, every frost, every stream contributing its share in the general waste. In the contemplation of such a history we are in a manner baffled and overawed by the vastness of the time which is required; the work accomplished is so vast, and the workers, even if we suppose that they once worked more rapidly and vigorously than they do now, seem so feeble. We may in imagination watch that ancient land for a thousand years, and yet detect no appreciable change upon its surface. We return to it after the lapse of a thousand centuries, and find perhaps that the valleys are only a little deeper, and that the broad undulations between them begin to bear but a far-off resemblance to hills. At the end of another long interval, during which perchance the land has undergone not a few unequal upheavals and depressions, the hollows have sunk a little more. But how impossible to realize, even if we may yet be able to estimate, the time which was needed to change the ancient table-lands into a region of mountains and valleys; to excavate the wide straths and glens; to scarp the cliffs and precipices; to roughen the mountain-sides with crag and scar and rocky pass; to dig the ravines and twilight gorges; and to carve out all that varied scenery which we know so well!

Perhaps it may yet be ascertained, that among the agents which in successive geological periods helped in no small degree to alter the surface of the globe were sheets of land-ice and fleets of floes and bergs. There are indications of old glacial periods in Palæozoic, Secondary, and Tertiary times, and we may eventually learn that the glaciers and bergs of some of these remote ages took a part in carving out the valleys and planing down the table-lands of this country. But, be that as it may, we know that the last great glacial period wrought marvellous changes upon the surface of the British Islands. In now glancing once more at the history of these changes, let us imagine the land, at the beginning of that period, rising into the same wide sweep of hill and valley as it does to-day. Gradually its plants and animals are displaced by those of more northern latitudes, as the temperature becomes year by year more wintry and ungenial. The snow creeps down from the hills, the forests and their inhabitants are pushed nearer and yet nearer to the sea; until at last, save perhaps in a narrow stripe along the shore in summer, one wide mantle of snow and ice has enveloped the land from the mountain peaks to the sea. Still the cold increases. The very ocean freezes into solid sheets around the shores. The high grounds of the interior—higher, perchance, by several thousand feet, than they are to-day—receive a constant accession of snow, and the accumulated mass, pressing down the valleys, goes out to sea in long wide walls of ice. As it descends, year after year, and century after century, the surface over which it moves is ground and polished, the hardest rocks are shorn down, the ruggedness of the ancient land is largely worn off, countless lake-basins are excavated in the rocks, and an undulating outline is impressed upon the whole length and breadth of the country. The moraine-rubbish of this great ice-sheet gathers into the thick deposit known as *boulder-clay*. The summer, brief and feeble, has yet strength enough to melt the last winter's snow along the coast and in the maritime valleys; and doubtless, under the fading skirts of winter, the bright flowers of an Arctic type—saxifrages, ranunculi, willows, mosses, and the rest—spring rapidly into bloom. Nor are the larger mammals wanting; in such sheltered nooks the mammoth and the rhinoceros would find their appropriate food, as their survivors, the rein-deer and the musk-ox, still do in the far north. The storms of summer work dire havoc on these shores, for the ground-swell, setting in strongly on the land, breaks up the coast-ice into heaps of ruin, which, laden with rocks and mud, are borne off, until they melt in mid-ocean or are stranded on other coasts.

The next act in this strange drama brings before us this ice-covered land slowly sinking beneath the sea. The higher mountain-tops, however, remain above water, and send out their fleets of bergs and ice-rafts, for the climate is still severe enough to nourish on the narrowed land an abundant growth of ice and snow. Many a huge mass of granite or gneiss or schist is thus dropt quietly over the coal-measures of the Lothians; many a block of grit and greywackè is borne from the lonely islets of Lammermuir, Moorfoot, Dumfries, and Galloway, and sunk upon the hills and valleys of the north of England. Nay, large boulders of mountain-limestone are lifted from the coast-line skirting the half-submerged hills of Northumberland and Yorkshire, and scattered far and wide over the central counties. Even from the distant shores of Scandinavia bergs bring fragments of gneiss and granite to the plains of Central and Southern England. To this and the later parts of the history the upper boulder-clay and sandy drift are probably to be assigned. Moreover, the grating of these ice-islands over the sunken hills must doubtless have given rise to much abrasion of the solid rocks. Many a Scottish hill-top may thus have been smoothed and striated anew; and we probably see proof of the same process in the large number of scratched fragments of chalk from the Yorkshire Wolds in the boulder-clays of the east of England. The land once more begins to rise; the glaciers in the mountains of Wales, Cumberland, the south of Scotland, and the Highlands resume not a little of their former massiveness, as the country gathers increase of size. The sand and clay which the sea may have left behind it are by the ice cleared out of the glens. With the widening area of land, and the lessening severity of the climate, the hills and valleys, where free from perennial snow, are clothed with vegetation and haunted by beasts of the chase. By a succession of changes, as slow and silent, doubtless, as those which ushered in the Age of Ice, that long era begins to draw to a close. The glaciers feel the breath of a warmer clime coming over them, and shrink step by step back into the mountains, leaving, at every pause in their decline, great heaps of earth and rock—memorials, as it were, of their final and fruitless conflict with the adverse elements. But their doom has come, and the last lingering remnant of the old ice-sheet vanishes away. The very plants and animals of that cold period are involved in the same fate. Slowly and reluctantly they are driven from the lower grounds, as species after species makes its appearance from other lands, like the successive hordes of a conquering people. And now at last, on the bleakest and barest of our uplands, from which there is no escape, they carry on the struggle still. But the skirmishers of the invading army are amongst them, and the time will doubtless come when the ancient and Alpine races will disappear from our highest mountain-top, and with them the last living terrestrial relic of the great glacial period.

Since the ice melted away, the sea, rains, streams, springs, and frosts have renewed their old work of demolition. The smoothed and flowing outline which the ice left behind it is now undergoing a slow destruction, and the rocks are quietly resuming the rugged outlines which they had of old. The sea-coasts are receding before the onward march of the waves; former ravines are deepened and widened by the rivers, and new ones have been formed. Man too has come upon the scene, and has set his mark upon well nigh every rood of the land from mountain-top to sea-shore. He has helped to demolish the ancient forests ; he has drained innumerable fens and mosses, and turned them into fertile fields; he has extirpated the wild beasts of the old woods, thus changing both the aspect of the country and the distribution of its plants and animals. He has engraved the country with thousands of roads and railways, strewn it with villages and hamlets, and dotted it with cities and towns. And thus more has been done by him, in altering the aspect of the island, than has been achieved during the brief period of his sojourn by all the geological agencies put together.

Such in outline is the explanation which I have ventured to propose for the origin of the present scenery of Scotland. It is based not upon speculations regarding former Titanic convulsions, of which we know nothing, but upon the order of nature as it has been ascertained by experience. Not that the brief recorded experience of man is to be taken as a standard by which the past eternity is in everything to be measured. But it is vain to feign causes which cannot be shown to have existed except by the evidence which they are invented to explain. It is not only safer, but it seems the only philosophical course, to interpret the past changes of the earth's history by constant reference to the well-ascertained economy of nature. And that the various processes now at work in altering the surface of the globe are enough to have given rise to all the varieties of scenery in these islands must, I believe, be admitted by every one who is not unwilling to allow the necessary lapse of time. But while it appears to me certain that our glens and valleys were excavated by the same agencies of denudation which are performing a similar task to-day, there is no proof that the rate of waste has always been the same. There may have been periods when rain and frost, for instance, were much more abundant and severe than they are now, and when in consequence the general waste of the land was more rapid. Geological evidence favours this supposition. Hence, to take the present rate of waste as the standard for all past time may be to fix the estimate much too low.

But, even with this limitation, we cannot contemplate the present landscapes of our country as the result of a necessarily slow and unequal decay without being impressed with a sense of the vastness of the time which is demanded. The astronomer may yet perhaps be able to calculate for us the length of some of these periods; but, even though they were expressed in figures, the mind would relinquish in utter despair the attempt adequately to realize them. In this dim, shadowy antiquity, so impressive from its immensity, and from the slow and stately march of the events which it witnessed, there is surely an ample equivalent for the grandeur of the Titanic upthrows of the older faith. It was doubtless pleasant to contrast the peace and quiet of the living world, with what seemed the records of fierce cataclysms in earlier times; to see the fair meadows and cornfields of the Lowlands sweeping in waves of verdure up to the base of crags that were believed to have been heaved into the air when the earth was shaking and tossing like a storm-vexed sea; to listen to the ripple of the river, and think that the tree-shaded ravine in which it flows was rent asunder by some primeval earthquake. But surely it is not less pleasant, nor less fraught with material for suggestive meditation, to wander away among the purple hills, and to meet everywhere with the proofs of slow change carried on during innumerable centuries; to trace the lines of scarred cliff and precipice with the chaos of fallen ruins below; to linger by the side of lake and glen; or to climb the heights above, that the eye may sweep over a wide sea of mountain-tops: and to believe that all this range of scenery has, inch by inch, been carved out by the joint labours of the powers of waste.

The story of the origin of our scenery, as thus interpreted, is of a piece with the rest of the teachings of nature. It tells of no sudden and random changes. It leads us back into the past infinitely further than imagination can follow, and, with an impressiveness which we sometime can hardly endure, points out the vastness of the antiquity of our globe. It shows that in the grander revolutions of the world, as well as in the humbler routine of every-day life, it is the little changes which by their cumulative effects bring about the greatest results; that the lowly offices of wind and rain, springs and frosts, snow and ice, trifling as they may appear, have nevertheless been chosen as instruments to mould the giant frame-work of the mountains; and that these seemingly feeble agents have yet been able, in the long lapse of still untold ages, to produce the widest diversity of scenery; and to do this, not with the havoc and ruin of earthquake and convulsion, but with a nicely balanced harmony and order, forming out of the very waste of the land a kindly soil, which bears, year by year, its mantle of green, yielding food to the beast of the field and the fowl of the air, and ministering to the wants and the enjoyment of man.

INDEX.

Abbey Craig, 277.
Aberdeenshire, Old Red Sandstone of, 110.
Aberlady, 49.
Agassiz, Professor L. cited, 160, 166, 195, 201.
Airdrie, 304.
Alluvial Meadows, or "Haughs," 29.
Annandale, 234, 249, 252.
Applecross, 104.
Arbroath, 56, 324.
Ardnamurchan Point, 104.
Ardrossan, 324.
Arisaig, Valley Road to Fort William, 124.
Argyle, Duke of, cited, 166.
Argyle's Bowling Green, 222.
Argyleshire, Hills of, 222.
    Valleys in, 138, 157.
Arthur's Seat, 277.
Assynt Mountains, 212, 214.
Atlantic, Breakers of, 42, 65, 67.
Atmospheric waste, 15—38, 40, 42, 114, 195, 209, 228, 264, 292, 317, 335.
Ayrshire Coast, 70, 72, 75, 234, 323, 324.
    Scenery of, 234, 262, 278, 297.

Banff, Old Red Sandstone of, 110.
Barnbougle (Linlithgowshire), 51.
Barra Head, 68.
Basalt, Mode of weathering of, 294.

Bass Rock, 269.
Beach, Formation of a, 42.
    Illustrates a river-system, 11.
"Beallochs," or passes, 123.
Beauly Firth, 59, 144.
Bell Rock, Force of waves at, 54.
Ben Aulder, Height of Spring on, 117.
    Griam, 109.
    Lawers, Structure of, 96.
    Lomond, 142, 223.
    Nevis, Ascent of, 116.
        Height of Spring on, 117.
        Glens on East Side of, 117, 123, 217.
        View from top of, 99.
Berwickshire Coast, 46, 232.
    Old Red Sandstone of, 246.
Biggar, 288.
Binny Craig, 269, 277, 294.
Black Isle, 132.
Borthwick Water, 253.
Boulders, Travelled, 190, 312, 346.
Boulder-clay, Composition of, 303.
    Fossils of, 304.
    Origin of, 185, 188, 257, 300, 345.
    Ravine of, 31, 36, 302.
    Scenery produced by, 185, 307.
    Subdivisions of, 301.
Brae Lyon, 142.
Braes of Doune, 110.
Brander, Pass of, 174, 175.
Breaker-action, 40, 60.
Brick-earth, 16.

A A

Index.

Broad Law (Peebleshire), 238.
Brooks, Action of, 18.
Brora, 203.
Burntisland, 52, 324.
Bute, 155, 270.
    Kyles of, 69, 155, 157, 222.
Button Ness, 53.

Cairn Gorm, 208, 223.
Caithness Coast, 60, 109.
    Ord of, 188.
Caledonian Canal, Valley of. *See* Great Glen.
Callander, 110.
Calton Hill, 295.
Cambrian Sandstones, Occurrence of, in N.W. Highlands, 93.
    Influence on scenery, 210.
Campsie Fells, 142, 269, 278, 297.
Carboniferous Rocks on Southern Uplands, 233, 247, 252.
    of Midland Valley, 267, 268, 270, 271.
Carnwath, 312.
Carrick, 235, 268, 316.
Carstairs, 311.
Cartland Craigs, 290.
Cataclysms, 8.
Chambers, Mr. R. cited, 160, 164, 260.
Cheviot Hills, 142.
Clyde, Firth of, 70, 155, 169, 222, 270, 300, 305, 323, 324.
    River, 148, 267, 284, 287, 305.
Coast Scenery, 41—76, 125, 169, 226, 232, 320.
Coll, Island of, 75.
Colorado, River of the West, 33.
Conical form of Mountains, Origin of, 214—217, 219, 292.
Contin River, 131.
Corstorphine Hill, 295.
Cove, Berwickshire, 47.
Cowal, 223.
"Crag and Tail," 295.
Crail, 52.

Crianlarich, 142.
Crichhope Linn, 265.
Croll, Mr. James, on Cause of Cold of Glacial Period, 166, note.
Cromarty Firth, 60, 131, 141, 187, 324.
    Sutors, 132.
Cuchullin Hills, Skye, 210.
Culbin, Ruined Barony of, 75.
Culross, 52.
Cumnock, 235.

Dalmellington, 235.
Dalveen, Pass of, 238.
Dee, River (Aberdeenshire), 145.
    (Galloway), 250.
Deltas in Scottish Rivers, 129.
Denudation, Results of, 85, 98, 106, 114—139, 149, 184, 212, 217, 244, 248, 262, 272.
Dirie More, The, 222.
Don, River, 145.
Dornoch Frith, 75, 131, 144.
Drift. *See* Boulder-clay.
Drift-wood under pitchstone, Eigg, 280.
Drumouchter, Pass of, 125, 202.
Dumfriesshire, Glens in, 238.
Dunbar, 48.
Dundee, 324.
Dunglass Bay, 47.
Dunoon, 222.
Dysart, 52.

Earthquakes, commonly supposed to explain many forms of scenery, 8, 21.
Earth's crust, its dislocated structure, 9.
Eathie, 177.
Edinburgh Castle Rock, 269, 294.
Edinburghshire, 234, 268, 271, 294.
Eigg, Isle of, 226 (fig.) 278.
Elgin, Old Red Sandstone of, 110.
Ericht, River, 145.
Erratic Blocks, 312.
Escarpments, Formation of, 36, 38.

Index.

Esk, River, 145.
Ettrick, 253.

Falkirk, Carse of, 287, 322.
Faults, Influence of, in determining the form of the surface, 97, 121, 140, 177, 252, 254, 270, 285, 291.
Felspathic Igneous Rocks, their influence on scenery, 292.
Fifeshire, 52, 75, 274, 278, 296.
Findhorn, 58.
    Valley of the, 138, 144.
Fiords, or Sea-lochs, 125, 144, 183.
Floods in rivers, 338.
Forbes, Principal, J. D. cited, 160.
Forests, Prehistoric, and Peat-mosses, 327.
Forfarshire Coast, 55, 322.
Fort George, 59, 130.
Forth, Firth of, 48, 287, 322.
    River, 283, 286.
Fortrose, 130.
Fort William, Valley between, and Arisaig, 124.
Foster, Mr. C. L. Neve (Geological Survey), cited, 17.
Frost, 37.

Galloway, 235, 242, 257, 261, 316.
Gareloch, 129, 156, 222, 270.
Garlton Hills, 269, 278.
Geological Structure, its connexion with scenery, 91, 95, 98, 219, 232, 254, 270. *See also under* Faults *and* Subterranean Movements.
Geological Contrasts, 169, 324.
German Ocean, Breakers of, 53—64.
    Bed of, 73.
Girvan, 234, 270.
    Valley of the, 235.
Glacial Period, Stages of the, 149, 166, 191, 205, 344.
    Close of, 205, 230, 264, 348.
    Cause of Cold of the, 166, note.

Glacial Periods, Successive, in the Geological Past, 344.
Glaciation of the Highlands, 123, note, 149—206.
    of the Midland Valley, 299—317.
    of the Southern Uplands, 256—265.
Glaciers, Action of, Chap. iv. 149, 159, 197.
Glasgow, 323, 324.
Glens, Origin of, Highland, 115.
Glen App, 253, 254.
    Collarig, 197.
    Darual, 157.
    Dochert, 146.
Glenelg, 219.
Glen Garry, 125, 146, 202.
    Glaster, 200.
    Gluoy, 200.
    Lyon, 96.
    Morriston, 180.
    Ogle, 147.
    Roy, 194, 197.
    Shiel, 143, 218.
    Spean, 167, 193.
    Treig, 195.
    Truim, 125, 202.
    Urquhart, 180.
Gneissose Rocks of Highlands, 93.
    Influence of, on Scenery, 218.
Goatfell, 223.
Golspie, 109.
Gowrie, Carse of, 322.
Grampians, 112, note, 113, 163, 224.
Granite, Occurrence of, 97.
    Influence of, on scenery, 208, 223.
    Mountain, Waste of a, 117.
Granton, 49.
Great Glen, 97, 109, 138, 140, 155, 177.
Greenland, Present Aspect of, like that of Scotland during the glacial period, 78, 161.
Greenock, 324.

Greenock, View of Highland Table-land from, 105.
Greenstone, Weathering of, 292.
Grey Mare's Tail, 238, 260.

Haddingtonshire, 234.
Hall, Sir James, cited, 295.
Harris, Isle of, 68.
Hartfell, 142, 238.
"Haughs," 30.
Hebrides, 67.
Helmsdale, 109.
Hibbert's *Shetland Islands* quoted, 65.
Highlands, Geological Structure of, 91, 108, 270.
    Denudation of, 98, 114, 122, 139, 149—185, 212, 217.
    Ancient Table-land of, 99.
    Geological age of table-land of, 108—111.
    Origin of Valleys of, 114.
    Glaciation of, 151.
    Origin of Lakes of, 171.
    Glaciers of, 180, 191, 201, 299.
    Higher Elevation of, during glacial period, 183, 188.
    Submergence of, during glacial period, 189.
    Local varieties of scenery in, 207.
    Precipices in, 224.
    Landscapes on borders of, 225.
    Atmospheric waste in, 119, 208, *et seq.*
Holy Loch, 223.
Hutton, 10.

Ice, Geological Effects of, 37, 77, *et seq.* 151, 159, 173, 181, 195, 256, 299.
Icebergs, 82, 161, 164, 173, 256, 304, 316, 344.
Ice-worn Rocks, 151, *et seq.* 228, 262, 299.
    Weathering of, 228, 264, 316.
Innellan, 222.

Invergordon, 131.
Inverness and Perth Road, 125, 202.
    Firth, 130.
Invernesshire, Mountains of, 218.
    Valleys in, 125, 138, 143, 218.
Isla, River, 145.
Islay, Isle of, 214.

Jamieson, Mr. T. F. cited, 167, 196.
John O'Groat's House, 144.
Jukes, Mr. J. B. cited, 141, note.
Jura, Island of, 214.
    Sound of, 156, 174.

Kames, 308.
Kane, Dr., on Greenland glaciers, 79.
Kilmartin, 175.
Kilpatrick Hills, 278.
Kincardineshire Coast, 57.
Kinlochewe, 143.
Kinloss, 58.
Kinnoul Hill, 284.
Kintail, 143.
Kirkcaldy, 52.
Kirkcudbright Coast, 71, 232.

Lakes, Origin of, 81, 171, 192.
Lammermuir Hills, 233, 234, 236, 250, 251, 270.
Landscape painting and geology, 216.
Land-surface, Sketch of the history of a, 87.
Lanarkshire, 234, 311.
Lauder, Sir T. D., on Morayshire Floods, cited, 32.
Laurentian or Fundamental Gneiss, Occurrence of, 93, 176.
    Its influence on scenery, 210.
Leader, Valley of, or Lauderdale, 233, 249, 250, 251.
Legends, Connexion of, with geological features, 3, 309, 313.
Leith, 49.

Lewis, Isle of, 176.
Limestone, Influence of, in Highland landscape, 217, 220.
Linlithgowshire, 269, 278.
Lochaber, 180, 193.
Loch Alsh, 143.
    Arkaig, 180.
    Aven, 220.
    Awe, 138, 174.
    Carron, 124, 129, 138, 143.
    Craignish, 144.
    Doon, 262.
    Duich, 143.
    Eck, 125.
    Eil, 180, 193.
    Etive, 128, 175.
    Fannich, 222.
    Fleet, 130.
    Fyne, 128, 129, 138, 140, 144, 156, 166, 183, 205, 223, 228.
    Garry, 202.
    Hourne, 219.
    Keeshorn, 143.
    Killisport, 144.
    Laggan, 142, 196.
    Leven (Kinross), 296.
    Linnhe, 138, 144, 155.
    Lochy, 142.
    Lomond, 128, 156, 222.
    Long, 144, 156, 166, 222.
    Maree, 128, 214.
    Morar, 128.
    Ness, 177, 183.
    Nevis, 219.
    Oich, 142, 178.
    Quoich, 142.
    Riden, 157.
    Ryan, 70, 252, 254.
    Shiel, 138.
    Skene, 259, 335.
    Swene, 144.
    Tarbert, 144.
    Tay, 96, 138, 140, 146.
    Torridon, 143, 214.
Lomonds of Fife, 296.
Lothian, East, 234, 236, 246, 268, 269.
Lothian, Mid, 234, 268, 269, 271.
    West, 269, 278.
Lowther Hills, 238.
Luce, Bay of, 252, 261.
Lyell, Sir C. cited, 57, 160.
Lyne Water, 250.

Maclaren, Mr. C. cited, 160, 167, 295, 320.
Macrihanish Bay, Cantyre, 75.
Maitland, *History of Edinburgh*, quoted, 49.
Mammoth, Bones of, in drift, 304.
Man, Aspect of Scotland at first coming of, 327.
    Influence of, upon scenery, 336.
    Changes of scenery since his advent, 319.
Marine denudation, Chap. III.
    Forms a platform or plain, 72, 106, 244.
Mathers, Kincardine, 57.
Merrick, 242, 256, 261.
Midland Valley, 266.
    Geological structure of, 266.
    Denudation of, 272, 291.
    River-systems of, 282.
    Local varieties of scenery in, 292.
    Prominence of igneous rocks in, 292.
    Tameness of stratified rocks in, 298.
    Glaciation of, 299—317.
Miller, Hugh, cited, 66, 75, 187.
Moffatdale, 238.
Monadhleadh Mountains, 142, 180.
Moncrieff Hill, 274, 284.
Mountains and Valleys, Popular ideas regarding, 1.
    Due to erosion of valleys, 10, 13.
Moraines, 81, 192.
Moray Firth, 58, 75, 92, 109, 155, 177, 187, 225.
Muck, Water of, 253.
Mull, Isle of, 226.

Index.

Murchison, Sir R. cited, 92, 94.
Musselburgh, 49.
Myths, Geological origin of, 3, 309, 313.

Nairn, Old Red Sandstone of, 110.
    River, 144.
Ness, River, 144.
Newhaven, 50.
Nithsdale, 234, 235, 249, 250, 252.
North Berwick, 49.
    Berwick Law, 269, 277, 293.
Norway, 162.

Ochil Hills, 190, 267, 278, 286, 293, 300, 315.
Old Red Sandstone of Highlands, 91, 108, 110, 225.
    Midland Valley, 267, 270.
    Southern Uplands, 233, 246, 249.
Oolitic strata, Influence of, in landscape, 225.
Orkney Islands, 62.

Parallel Roads of Glen Roy, 197.
Peach, Mr. C. W. On force of waves along Caithness Coast, 60.
    Shetland Islands, 64.
Peat-mosses, 327.
    Decay of, 334.
Peat, Encroachments of, on lakes, 311, 330.
Peeblesshire, 234, 238, 258.
Pennant's *First Tour* quoted, 76.
Pentland Firth, 61, 155.
    Hills, 142, 268, 272, 293.
Permian rocks of Southern Uplands, 234, 247, 252.
Perth and Inverness road, 125, 202.
Pitchstone-porphyry of Eigg, 279.
Playfair, 10.

Pot-holes in bed of a river, 28.
    formed by sea on a beach, 73.
Precipices of the Highlands, 224.

Quartz-rock, Occurrence of, in Highlands, 93.
    Scenery of, 214.
Queensferry, 52.

Rain, 15, 139.
"Rain-wash," 16.
Raised Beaches, 193, 318.
Ramsay, Professor A. C. cited, 17, 38, 81, 107, 112, 115, 126, 172, 303.
Rannoch, Moor of, 97, 142, 223.
Ravines, Popular view of their origin, 21.
    Geological history of, 22, 32.
Reindeer, Bones of, in drift, 304.
Renfrewshire, 268, 297.
River-action, 20, 34.
    Systems, their nice adjustment, 6.
    Valley, Playfair's description of a, 10.
    Terraces, 30.
Rivers of Highlands, 137.
    Midland Valley, 282.
    Southern Uplands, 249.
    Scottish, Diminution of water in, 337.
Roan Island, 109.
"Rock-basins" in glaciated countries, 81, 128.
Rosemarkie, 188.
Ross, Easter, Plains of, 132.
Ross-shire, Coast of, 210.
    Geology of, 92, 109.
    Mountains of, 210.
    Valleys in, 124, 138, 143.
Rothesay, 324.
Roy, Bridge of, 194.

St. Abb's Head, 46.
    Andrew's, 52, 75, 320.
    David's, 52.

Index. 359

Salisbury Crags, 276, 295.
Salmon in rivers above high falls, 288, note.
Sand-spits, 129, 130.
Scenery, Prevalent uncertainty regarding origin of, 5.
Schehallien, 214.
Schistose rocks of Highlands, Influence of, on scenery, 208.
Scottish scenery, Types of, 89.
Sea, as a geological agent, 39 *et seq.*
Sea-lochs, 125, 144, 183.
Shandwick, 134, 177.
Shee, River, 145.
Sheep-drains converted into gullies by rain, 19.
Shells, Northern, in deeper parts of Scottish seas, 205, 306.
   in glacial drift, 305.
Shetland Islands, 63.
Sidlaw Hills, 267, 274, 278, 284, 300, 315.
Silurian Metamorphic Rocks of Highlands, 92—95.
   Their scenery, 208, 214—225.
   Rocks of Southern Uplands, 231.
   Influence of, on scenery, 234—244.
Skateraw, 48.
Skerries and reefs, Production of, 42.
Skerryvore Rock, Breakers at, 67.
Skye, Hills of, 208, 210, 223, 226.
   Kyles of, 205.
Smith, Mr. James, of Jordanhill, 305, 320.
Spey, River, 138, 144.
Springs, 34.
Stevenson, Mr. quoted, 49 *et seq.*
Stincher, Valley of, 253, 254.
Stirling, 286, 302.
   Castle Rock, 269, 277, 294.
Stranraer, 71.
Strath Brora, 203.
   Spey, 200.
Stroma, 62.

Submergence of Scotland during glacial period, 188, 190, 258, 287, 316.
Subterranean movements, Influence of, in determining the configuration of a country, 86, 121, 133, 140, 146, 270, 282, note, 323. (*See also under* Earthquakes *and* Faults.)
Sutherlandshire, Coast of, 203, 210, 225.
   Geology of, 92, 108.
   Mountains of, 210.
   Valleys of, 138, 143.

Tain, 132, 188.
Talla, Valley of, 238, 240, 258.
Tarbat Ness, 132.
Tay, Firth of, 53, 75, 274, 283.
   River, 145, 284.
Tent's Muir, 75.
Teviot, 253.
Till, or boulder-clay, 185.
Time, its value in geological history, 8, 121, 350.
Tinto, 288, 293.
Tiree, Island of, 75.
Topley, Mr. (Geological Survey), cited, 17.
Traprain Law, 293.
Trap-rocks, Influence of on scenery, 208, 215, 226, 267, 268, 274, 292.
Trosachs, 208, 218.
Tweed, River, 46, 148, 253, 288.
Tyne, 49.

Uplands of South Scotland, 231.
   Geology of, 231.
   Scenery of, 234, 255.
   Surface of, an old sea-bottom, 244.
   Age of table-land of, 246.
   Valleys of, 248.
   Ancient glaciers and icebergs of, 256, 300.
   Recent atmospheric waste in, 264.

Valleys due to atmospheric waste, 10, 34.
  or passes across watersheds, 123.
  of Highlands, Origin of, 114, 137, 139.
  Longitudinal, in Highlands, 138.
  Longitudinal, in Southern Uplands, 249, 253.
  Transverse, in Highlands, 138.
  Transverse, in Southern Uplands, 249, 254.
Victoria Falls of River Zambesi, 33.

Wales, Submergence of, during glacial period, 189.

Water, Circulation of, upon the globe, 14.
Waterfalls, Geological history of, 26.
Watershed of Scotland, 142.
Waves, Force of, 40, 54, 60, 64, 67, 71.
  Their inability to rub down the surface of rocks, 41.
Wick, 60, 188.
Wigtown Coast, 71, 232, 234, 323.
Wind, Geological effects of, 74.
Wrath, Cape, 142.

Yarrow, River, 253.
Young, Dr. John, F.R.S.E. cited, 260, 278.

THE END.

R. CLAY SON, AND TAYLOR, PRINTERS, BREAD STREET HILL LONDON.

SD - #0011 - 020725 - C0 - 229/152/21 - PB - 9781330493403 - Gloss Lamination